Algorithms Are Not Enough

Algorithms Are Not Enough

Creating General Artificial Intelligence

Herbert L. Roitblat

The MIT Press
Cambridge, Massachusetts
London, England

This book was set in ITC Stone Serif Std and ITC Stone Sans Std by New Best-set Typesetters Ltd. Printed and bound in the United States of America.

Library of Congress Cataloging-in-Publication Data

Names: Roitblat, H. L., author.
Title: Algorithms are not enough : creating general artificial intelligence / Herbert L. Roitblat.
Description: Cambridge, Massachusetts : The MIT Press, [2020] | Includes bibliographical references and index.
Identifiers: LCCN 2019046398 | ISBN 9780262044127 (hardback)
Subjects: LCSH: Artificial intelligence.
Classification: LCC Q335 .R65 2020 | DDC 006.3—dc23
LC record available at https://lccn.loc.gov/2019046398

10 9 8 7 6 5 4 3 2 1

Contents

Preface

At least since the 1950s, the idea that it would be possible to soon create a machine that was capable of matching the full scope and level of achievement of human intelligence has been greeted with equal amounts of hype and hysteria. We have now succeeded in creating machines that can solve specific fairly narrow problems with accuracies that meet or exceed those of their human counterparts, but general intelligence continues to elude us. In this book, I want to outline what I think it will take to achieve not just task-specific intelligence, but general intelligence.

Although some people look forward to achieving artificial general intelligence, others fear it, to the point of predicting that a generally intelligent machine will spell the end of human existence. Such a machine would be able to improve itself, their thinking goes, and will quickly pass from equaling human intelligence to far exceeding it. Computers will become so intelligent that humans will be lucky to be kept as pets. At best, the intelligent computers will ignore us; at worst, they will seek to destroy us as pests competing for resources.

Both views are fundamentally untenable. The tools that let us build specialized intelligence are not up to the task of general intelligence. Even if we make new tools that are capable of achieving general intelligence, they will not result in any kind of explosive self-improvement in intelligence. I describe why improvements in machine intelligence will not lead to runaway machine-led revolutions. Improvements in machine intelligence may change the kind of jobs that people do, but they will not spell the end of human existence. There will be no robo-apocalypse.

I have written this book for a nontechnical reader. If I succeeded, you should not have to know much about computers, psychology, or artificial intelligence to read it.

Read this book if you are interested in intelligence, if you want to know more about how to build autonomous machines, or if you are concerned that these machines will someday take over the world in a sudden explosion of technology called "the technological singularity." Hint: they won't.

I hope to convince you that it is possible to create artificial general intelligence, but it is neither so imminent nor so dangerous as some authors would have you believe. It will take a change in perspective, and I have tried to sketch out just what that new perspective is.

This topic is important because hardly a day goes by without a call for some kind of regulation of artificial intelligence, either because it is too stupid (for example, face recognition) or imminently too intelligent to be trusted. Although this is not a book about policy, good policy requires a realistic view of what the actual capabilities of computers are and what they have the potential to become. Conversely, progress in developing artificial general intelligence requires knowledge that we do not have about the nature of intelligence, brains, and the kinds of problems a generally intelligent agent will have to solve.

As Alan Turing said in 1950, "We can only see a short distance ahead, but we can see plenty there that needs to be done."

1 Introduction: Intelligence, Artificial and Natural

This chapter provides an overview of the book. It points out that artificial intelligence is not new; people have been inventing intelligence tools for at least 50,000 years. What is new, is running these methods in computers. When we use the word "intelligence," we typically mean the kind of higher cognitive functioning that we learn in school, but that kind of intellectual achievement rests on an existing foundation of natural intelligence. General intelligence requires an integration of the two.

When we talk about intelligence, we usually mean the kind of higher intellectual functioning that we learn in school. We mean things like logic and reasoning. At its height we mean the kind of thinking that earned Albert Einstein the Nobel Prize in Physics.

When we talk about artificial intelligence, we typically mean processes executed on a computer. The term "artificial intelligence" is usually attributed to John McCarthy, who used it in a proposal for a 1956 summer workshop at Dartmouth College on making computers emulate human intellectual functioning. But artificial intelligence is more general than that. It is an organized systematic approach to processing information. It does not matter whether these processes are executed on a machine, on paper, or in a brain. One of these processes, algebra, for example, allows people to think systematically and to solve mathematical problems that were previously intractable.

The invention of systematic processes has guided the development of human intelligence for at least the last 50,000 years. These processes are just as artificial as computers or spaceships—they were all invented by people. They are precisely what has allowed human technological processes to advance and thrive over that period.

On the other hand, it would be wrong to claim that intelligence consists exclusively of these higher intellectual functions. Intelligence requires more than that. Einstein was not recognized as brilliant for his ability to systematically solve mathematical equations, but rather for his ability to create new ideas, new views of the world that were captured in his equations. For example, the equation for which he is best known, $E = MC^2$, is extremely simple and almost trivial to solve, yet it embodies an idea that is utterly elegant, and one that continues to play an important role in theoretical physics. The main idea of this equation is that the relationship between energy and matter is invariant, despite the obviously different forms each can take.

Einstein's brilliance was not just a logical recombination of the work that had gone before, but was a leap beyond. He did not just deduce the physics principles from observations that had been made but instead predicted observations that would be made. His work transcended the facts that he knew about and predicted new facts. Human intelligence, including Einstein's, requires both a logical kind of systematic thinking and a nonlogical kind of thinking of the sort that allows insight.

We don't have a good vocabulary for talking about these complementary abilities. Roughly, we might talk about intuition, on the one hand, and deliberation on the other. We might talk about natural intelligence, which everyone might be able to achieve, and artificial intelligence, which requires education and training. We might talk about biological versus computational intelligence. Another Nobel Prize–winning scientist, Daniel Kahneman, talks about thinking fast and thinking slow.

Natural human intelligence is what allows babies to learn their mothers' face within a few hours of birth. It's what allows us to walk across a crowded room or fold laundry. Real natural human intelligence is nonrational, emotional, partly subsymbolic. It jumps to conclusions based on very little evidence.

Invented artificial intelligence, in contrast, allows adults to reason rationally about complex ideas in nonemotional ways. Invented artificial intelligence is rational, methodical, and symbolic, but it, too, has limitations. Artificial intelligence provides tools that allow people to reason carefully, to keep track of symbolic information, and to solve advanced problems in quantum physics, among others.

From a logical point of view, natural intelligence takes shortcuts. Natural intelligence is the source of many human foibles and quirks, but it also allows humans to respond to a dynamically changing world without getting lost in thought.

According to John McCarthy's proposal, along with Marvin L. Minsky, Nathaniel Rochester, and Claude Shannon, the goal of the Dartmouth Summer Workshop was to conduct a study toward the creation of a general artificial intelligence that would be able to form abstractions, solve problems, and improve itself. They thought, at the time, that the way to achieve this general intelligence was to describe as precisely as possible the nature of thought and get a machine to simulate it.

According to the participants, the workshop fell short of its lofty goals, but it can still be described as a profound milestone for the field of artificial intelligence. It is also telling that even at this early date, they focused on the kind of tasks that we associate with higher cognitive function. The participants viewed intelligence as rational, deliberate, and goal directed. For example, Allen Newell, John Clifford Shaw, and Herbert Simon were working on a program to prove mathematical theorems. Their Logic Theorist was intended to mimic the problem-solving skills of an adult human being—in this case, an expert mathematician. Their program would eventually prove 38 of the first 52 theorems from chapter 2 of Alfred North Whitehead and Bertrand Russell's book (*Principia Mathematica*). Some of the Logic Theorist proofs were even novel ones.

Herbert Simon is quoted telling a group of graduate students that he and Allen Newell, had over Christmas, "invented a computer program capable of thinking non-numerically, and thereby solved the venerable mind-body problem, explaining how a system composed of matter can have the properties of mind." Their choice of theorem-proving as their demonstration of mind within a computer was fortunate in that the process of theorem proving was already well-defined as a step-by-step process consisting of a small set of actions (for example, symbol substitution) that could be applied to a small set of basic facts or axioms (for example, symbols). The book that they imitated, in fact, was dedicated to proving the basic properties of mathematics, so it largely laid out the axioms and the operations that could be applied to those axioms.

In hindsight, Newell, Shaw, and Simon's work on the Logic Theorist was a small step from the symbolic logic of *Principia*, but at the time, it was a huge leap for computational intelligence. Their approach would have a profound effect on much of the work that came after it for many years. Even though Whitehead and Russell had laid out the steps for proving their theorems, it is instructive that the Logic Theorist did not always follow their methods. It proved some of the theorems in novel ways. Simon and his colleagues overestimated the importance of that finding, which was also a milestone in the development of computational intelligence, a tendency that is still commonly repeated.

Today we have computer systems that can play games, diagnose disease, and perform other tasks at suprahuman levels. Each breakthrough achievement is heralded as the next step in the evolution of computational intelligence, allegedly bringing systems closer to the goal of general artificial intelligence. If only we had a bit more memory and faster processors, we would at last be able to achieve general intelligence.

Many things have changed over the years since these early developments, but two things have not changed. One is the overreliance on a small set of processes as the necessary and sufficient ones to build a general intelligence. The computers of the 1950s and 1960s were far too slow and too limited to actually produce a full intelligence, so the researchers settled for solving example or "toy" problems. Their mistake lay in thinking that size and speed were the only limits to expanding these systems to fully achieving a humanlike intelligence.

Their other mistake was the belief that the kinds of problems that they were studying were fully representative of the kinds of problems that a general intelligence would have to solve. They focused on toy versions of problems with specifiable steps that are relatively easy to describe and specific solutions that are easy to evaluate. These kinds of problems can be described as "path problems." Solving them requires finding a path through a "space" that consists of all of the "moves" the system could make. Some combination of moves will solve the problem, and the computer's task is to find the specific path through the available moves that does actually solve it. Computational intelligence is the process of finding the set of operations and their order (the path) necessary to solve a problem.

Another way of describing these problems is, in the words of Judea Pearl, as exercises in curve fitting. To paraphrase his view, solving these problems consists of finding a function that maps the available inputs to the desired outputs. It is just a way of formulating statistical predictions. This mapping process can be quite complex, and the number of choices or estimates that go into forming that relationship can be daunting, but that is still the form taken by all of the current computational intelligence systems out there. But not all problems are like this. Not all problems are path problems.

The progress that has been achieved in computational intelligence, and it has been dramatic, has come from the genius of system designers to formulate systems that are within the capacity of computers to solve. These systems need not, and generally do not, perform the tasks in the same way that people do because computer scientists have figured out how to reduce them to Pearl's kind of estimation task. They may perform specific tasks better than people do, but this is not because they have exceeded human intelligence in that task but because their designers have found other ways to solve those problems that do not require humanlike intelligence. Maytag dishwashers may clean dishes cleaner than I do by hand, but that does not make them any closer to achieving the intelligence of a human restaurant employee.

None of this is to say that machine learning systems that diagnose disease, understand speech, or drive cars are not intelligent, but they are intelligent in a special-purpose way, not in a general way. If we are to get beyond special-purpose intelligence, we will need to solve problems that are not being addressed today. If we want humanlike intelligence, we must figure out a way to construct it from the tools that we have available or we must build new tools. There are some attempts to create general intelligence with current tools, but none of them, so far, has demonstrated any success. Rather, the more promising road is to try to understand and emulate how the only example of general intelligence we have, people, create this intelligence. Ultimately, machine general intelligence may not resemble human general intelligence in its specific methods, but it must resemble it in the range of its capabilities.

The Invention of Human Intelligence

Over thousands of years, we humans have invented ever more complex artificial thinking tools, but natural human intelligence does not seem to have changed much. To the extent that we are more intelligent than our Paleolithic ancestors, it is because we have combined natural intelligence like what they had with artificial intelligence invented over the centuries.

The inventions of language and then eventually writing were probably among the most important tools added to the human intellectual toolbox.

Although some people argue that language is somehow innate, it appears to have emerged somewhere between 100,000 and 50,000 years ago and to have profoundly expanded the capabilities of hominids (Gabora, 2007). Brains with language, as opposed to the same brains without language, have increased capacities to share information, to coordinate activity, and to transfer experience, among others (Clark, 1998). Language, and particularly syntax, was associated with an enormous expansion of the kind of cognitive processes that these early humans could engage.

According to William Calvin, "Words are tools." Calvin goes on to speculate that the prelanguage human may have been capable of words, which could be used in short expressions, but not capable of complex sentences or of talking about the future or the past. These humans may have been capable of some basic kinds of thought, but not capable of structuring those thoughts, and therefore not capable of manipulating images, hypotheses, or possibilities. Since the invention of language, human intellectual capabilities have changed substantially.

Modern humans migrated to Europe about 43,000 years ago. Cave paintings and carved figures from that period (33,000 to 43,000 years ago), along with musical instruments, were found in the Swabian Jura in southern Germany. The Paleolithic cave paintings in Chauvet cave near France's Ardeche River are thought to be 32,000 years old. According to some anthropologists, the structure and detail of these cave paintings imply that the painters enjoyed a relatively sophisticated mental world. The Lascaux paintings in southwestern France are only about 20,000 years old. During this period, humans began to bury their dead, to create clothes, and to develop complex hunting strategies, such as using pit traps to capture prey. In Asia, cave paintings from the Indonesian island Sulawesi are thought to date from about 35,000 years ago. On the Island of Borneo, figurative cave paintings have recently been described that appear to date from about 40,000 years ago. The cave paintings are an indication that the Paleolithic people were capable of symbolic representation of their environments.

Few artifacts of Paleolithic artificial intelligence survive, but among these are structures that appear to be symbolic of their builders' world. These artifacts may have played a role in helping people navigate their world geographically and perhaps spiritually. Some of them, for example, depict constellations that would have been important to navigation. The painters of cave paintings may have believed that depicting such things as deer and bison would make it easier to hunt those prey.

There is some evidence that Mesolithic (the period starting about 11,000 years ago) people also developed artifacts that are more recognizably computational, such as calendars. Calendars are clearly important to agriculture, but they may also be important to hunter-gatherers—for example, to time the migration of birds and animals or to collect ripe fruits from distant locations that could not be observed directly.

These calendars used notched stones or bones, for example, to notate the passage of astronomical objects, particularly the moon. Larger structures, like Stonehenge in southern England (5,000 years ago), or an even older calendar structure found in Aberdeenshire in Scotland (about 10,000 years ago) were also astronomical calculators. The Aberdeenshire calendar consists of a series of pits dug in the shapes of the moon's phases, arranged in a 164-foot arc. The arc was aligned with a notch in the landscape where the sun would have risen during the winter solstice, allowing the lunar calendar to be corrected each year to match the solar year.

A Neolithic calendar, Newgrange, is in the Boyne Valley, County Meath, of Ireland. Built over 5,000 years ago, it marks the winter solstice using a roof box that allows sunlight to illuminate a buried chamber around the winter solstice.

Humans have gone from painting on cave walls to inventing interplanetary spacecraft because they have, over many generations, developed thinking tools that enable increasingly sophisticated intellectual activity. Among these tools are:

- mathematics (starting about 4,000 years ago)
- logic (about 2,600 years ago)
- algorithms (about 800 years ago)
- digital computers (about 80 years ago)

Each of these inventions enabled many other inventions and discoveries, which further contributed to human intelligence. Without these tools, human thought tends to be incomplete, irrational, and biased. People jump to conclusions based on wishful thinking and incomplete information. Decisions are made on the basis of how easy it is to think of answers rather than on the correctness of those answers.

The invented thinking tools, on the other hand, make it possible for people to reason systematically and effectively. Being invented, they are artificial tools that enable human intellectual achievement. Our formal education system is designed to provide the artificial intelligence tools needed to power human intellectual achievement—for example, by training people in the use of logic.

Humans have become more intelligent by incorporating thinking artifacts into their process, by becoming more artificial. And those aids are necessary, because without them, human intellectual capabilities are limited. The aids themselves do not have to be particularly intelligent, but they work to make human thought more systematic and rigorous.

Computational Intelligence

For the last 60 years or so, computer scientists have been pursuing ways to make computers intelligent and predicting that computational general intelligence is imminent. By saying that they aspire to computational general intelligence, they typically mean that they aspire to creating a computer system that demonstrates the same intellectual capacity as a human can—or even better. Until relatively recently, much of this effort has been focused on emulating the systematic intellectual tools that I have labeled artificial intelligence without paying much attention to the other characteristics I have mentioned.

Herbert Simon claimed in 1956 that general computer intelligence was 10 to 20 years into the future. Mark Zuckerberg claimed in 2016 that computational intelligence will be available in 5 to 10 years. Contrary to repeated optimistic predictions, however, computational intelligence has achieved a number of specialized capabilities, but general intelligence continues to be elusive.

The failure to deliver on the promise of general intelligence is due, in part, to a focus on a restricted set of tasks without recognizing that the totality of human achievements rests on a foundation of more fundamental biological skills (such as perception). There are many reasons for this focus, but a key one is the implicit belief that intelligence consists mainly of the kind of skills that are shown when playing chess or diagnosing disease. This approach implicitly assumes that deliberation is the key function of intelligence and that all kinds of intelligence can be reduced to such skills.

These deliberative skills are the capabilities that amplify human intelligence, but they alone are not enough to achieve general intelligence. General intelligence involves more than special-purpose algorithms for individual tasks. The specific algorithms succeed precisely because they reduce an otherwise complex problem to simpler problems that can be solved by calculation. The inventiveness on which they depend is provided by humans.

Natural Intelligence

Until recently, social science theory, particularly in economics, has viewed people as fundamentally rational. Rational people are consistently deliberative, thoughtful, and self-interested. They pursue their goals in the most efficient manner possible. A rational actor seeks to achieve the highest possible well-being given the available information, opportunities, and constraints. In short, greed is good, and we can rely on people to be greedy. Thought, in this view, is based solidly on logical self-interest.

When a person does not act rationally, in this view, in his or her self-interest, it is because the person's thought processes, the person's logical judgment, has been contaminated by emotions. Nonrational choices are mistakes and not predictable as a guide to the person's true actions. Unfortunately, this strategy has also permeated a lot of thinking in artificial intelligence research.

Newell, Simon, and Shaw's Logic Theorist was purely logical, starting with axioms (basic irreducible logical assumptions) and operations and ending with logical proofs. In fact, Newell and Simon argued for the physical symbol system hypothesis, according to which a physical system that manipulated symbols was both necessary and sufficient to produce intelligence.

In contrast to this approach, the so-called Moravec's paradox (it's not a paradox at all, but that is what it is called) notes that it is relatively easy to get computers to execute high-level reasoning, but extremely difficult to achieve computational versions of the skills of a two-year-old child. Deliberative skills are easy to describe and easy to implement with a computer, but creating a computer that can walk across a crowded room, or a robot that can fold laundry, remains a challenge.

The explosion of interest in computational neural networks in the 1980s and 1990s made some headway in solving some of these challenges by adopting a more biologically inspired approach to artificial intelligence. Instead of high-level deliberative rules, neural networks employ models that are more like simplified neurons. Instead of operating on symbols, like the words in a language, neural networks use connections among simulated neurons. The widespread use of neural networks, which have now grown into so-called deep learning models, was responsible for a lot of progress in computational intelligence, but it still did not bring us any closer to achieving general intelligence. Neural networks and other forms of machine learning helped to make it more obvious that the practice of AI, as opposed to the aspirations of AI, was complex functions that mapped inputs to outputs. As Hans Moravec and others asserted, it takes a lot more computation to simulate even a simple neural network than to follow a collection of rules, but both of them are still just calculating functions, an opinion shared by Pearl.

The key part of natural intelligence is the apparent ability to construct problem spaces, not just find paths through one that has already been constructed. But natural intelligence also has other properties. Natural intelligence is not concerned with finding the optimal solution to problems. Rather, natural intelligence is willing to jump to conclusions that cannot be "proven" to be correct in any sense of the word.

Rather than being algorithmic as artificial intelligence is, natural intelligence is heuristic. An algorithm is a set of steps that when followed with a particular input will always yield a corresponding output. A heuristic, on the other hand, is more like a rule of thumb. It mostly works, but sometimes it does not. A baby can recognize his or her mother within hours after birth, but a computer learning to identify categories of objects may require several thousand presentations. Take a child to the zoo and buy him cotton candy, and that kid will expect the same treat on all future visits.

In contrast to the intellectual capacities modeled by computational intelligence, many of the basic cognitive functions that I have called natural intelligence are shared by other species. Precocial birds (birds that can feed themselves immediately after hatching), such as chickens and ducks, learn to identify their parents within hours of birth. Scrub jays and other birds can store seeds under rocks and in crevices and recover them even months later after their environment has been covered by snow. As Wolfgang Köhler showed, chimpanzees can solve certain kinds of insight problems. Rather than learn by trial and error, chimpanzees were observed to put two sticks together or to stack boxes in order to reach food that was otherwise out of their reach.

Many animals, from ants to bears and chimpanzees, have been found to be able to respond to small numerical quantities (typically on the order of one to four or six) when other features have been controlled. Dogs and other animals can learn the names of up to about a thousand objects with some training and can select those objects following verbal commands.

Natural human intelligence or that found in animals can play an important role in that species' cognition. But the full intellectual achievement of humans up to this point has depended on using that native intelligence plus additional thinking tools that have been invented to achieve the current level of intellectual functioning.

Human natural intelligence has mostly been studied in the context of the foibles and failures it produces in educated humans or in the context of psychological development. It has been largely neglected as a source of human achievement, so we know a lot about the biases and limits it imposes on intelligence, but little about the positive contributions it makes. Natural intelligence is extremely likely to play a critical positive role in general human intelligence, and if we can figure it out, likely to play an important role in computational intelligence as well. Humans could not have invented their thinking tools without it and could not function if they were limited to trial-and-error learning as the early psychologists argued, or to the repeated presentation of labeled examples as modern machine learning would suggest.

The General in General Intelligence

Just how general does general intelligence have to be?

Einstein was really successful at theoretical physics. He won the Nobel Prize for his work on the photoelectric effect—which is the basis for how solar cells generate electricity. Arguably, his work on relativity was even more impactful. As smart as he was, though, Einstein was not good at everything. He was not, apparently, a distinguished mathematician, though he used mathematics very effectively. He may have played chess, but he is very unlikely to have been an accomplished go player. I doubt that he would have done well on the television game show *Jeopardy!*.

There are clear differences among people in their ability to learn, understand, create, analyze, interpret, and adapt to their environments. But not all of these abilities are equal. Einstein could play the piano and the violin, but it was doubtful that his skill with these instruments would have compared favorably to that of Itzhak Perlman or Mozart. Yo-Yo Ma is a great cellist, but I don't think that he has any publications in physics journals. Intellectual performance can vary from task to task, from time to time, as well as from person to person. Although there may be correlations among a person's capability on different skills, that is, a person who performs well on some task is likely to perform well on some others (See chapter 2), being brilliant on some tasks does not guarantee that you are brilliant on others.

Intelligence is a complex concept that involves many different kinds of skills. Psychologists have been measuring intelligence for over a century, but they are mainly interested in identifying the differences among people, rather than identifying the mechanisms by which it is produced. The first intelligence tests were designed to detect students who might need special help in school. The goal was to predict the overall aptitude of the person for learning or for other measures of intellectual success. Intelligence tests may include vocabulary assessments, analogies, image manipulation, or reasoning. Each of these has been found to correlate with some measures of success.

Intelligence tests usually include a battery of different subtests, each directed at measuring a specific ability. The idea of general intelligence as a thing comes from the observation that people's performance on these subtests tend to be correlated. If a person does well on a test that requires image rotation, for example, that person is likely to also do well at answering vocabulary questions.

This correlation among subtest performances has been called the "g-factor" for general intelligence. G could indicate the presence of some kind of general intelligence, for example, some people might have more powerful brains than others and so perform well. Alternatively, g may be merely a label for the statistical correlation. Intelligence, in other words, may not actually be all that general; instead it could be that the tests are not that good at isolating specific abilities. Multiple subtests may assess overlapping sets of specialized capabilities.

For example, a test taker who had vision problems might perform poorly over many tests not because that person is dumber than one with better vision, but because he has trouble reading the questions. People who are anxious might perform poorly on all tests, and those who are calm might perform better on all tests. Test taking may be its own skill. These associated factors may cause correlations without saying anything about general intelligence.

The correlations on the subtests of an intelligence test are not necessarily indicative or performance on real-world activities. Consider the relative skill sets of Albert Einstein and Yo-Yo Ma. Both are brilliant and are successful in their own, nonoverlapping ways. Intellectual superiority in one area does not guarantee superiority in other areas. We will consider the nature of the correlations in the context of intelligence tests in the next chapter. If human intelligence is any kind of example, artificial general intelligence may not, in the end, be quite as general as some people might expect.

Specialized, General, and Superintelligence

Computational intelligence programs so far have mostly involved performance on a single task, such as playing chess, diagnosing brain injuries, answering *Jeopardy!* questions, and the like. Chess playing was once thought to be a prime example of human intellectual capabilities. Chess was thought to be indicative of using strategy, reading the motivations of other people, and engaging in deep analysis of the situation. In this light, solving the problem of playing chess would go a long way toward addressing general intelligence because it would require the solution of so many higher cognitive functions. A chess-playing computer would have to assess its opponent, understand the person's motivations, and analyze the situation.

In fact, in his famous book, *Gödel Escher Bach*, Douglas Hofstadter argued that "there may be programs that beat anyone at chess, but they will not be exclusively chess programs. They will be programs of general intelligence, and they will be just as temperamental as people. "'Do you want to play chess?' No, I'm bored with chess. Let's talk about poetry" (Hofstadter, 1979, 1999, p. 678).

Instead, just the opposite happened. We have computer programs that are able to play chess at a very high levels, but they are incapable of also talking about poetry. The way chess-playing programs have been designed has nothing to do with deep psychological functions or general intelligence.

Rather, these programs depend on a simpler special-purpose method that organizes potential chess moves into a kind of branching tree. Algorithms are available to search among these branches and identify moves that are likely to lead to a successful outcome for the game. Chess developers reduced the problem of choosing chess moves to the simpler problem of selecting from a series of tree branches.

Playing the game go was predicted to be beyond the capacity of computers. Even the kind of approach that was successful for chess would not work for go, because of the huge number of different possible go positions and the number of ways they could be combined make the go tree too complex to evaluate moves in the same way they can be evaluated for chess. However, computer scientists were recently able to build a system that could play go at a world-class level, because they built another special-purpose algorithm.

The knowledge that went into developing programs that play chess or go is valuable for what it tells us about solving other similarly structured problems. Go became possible when the DeepMind team, who developed the program, designed useful heuristics to limit the number of branches that had to be evaluated to choose a move.

Given the reductionist approach to special-purpose computational intelligence, it should not be surprising that computers have not, so far, made much progress in general intelligence. The creation of yet another special-purpose algorithm may be intelligent, but even a collection of every special-purpose algorithm will not get us to a general intelligence.

Computer science has been effective at building hedgehogs, but not yet at building foxes. The ancient Greek poet Archilochus is commonly quoted as saying, "The fox knows many things, but a hedgehog one important thing." Current computational intelligence systems excel at specific tasks, but none of them yet has achieved any level of generality. There is no reason to think that combining special-purpose systems will, even eventually, result in the emergence of a general intelligence. A fox cannot be constructed from a stack of hedgehogs.

General intelligence, even in humans, is an elusive topic. The correlation among subtests could be due to some kind of brain efficiency, but it could also be a purely statistical artifact. If Einstein had a better brain, then maybe he should have been able to do everything better than other people, but his talent was limited. Intelligence in people, as measured by their successes and failures and not by their performance on tests, depends strongly on having certain kinds of experiences necessary to build expertise.

Even if brain efficiency is not the cause of superior human intelligence, it still could be a factor in improving machine intelligence. Computers get faster every year and processes that were impractically slow a few years ago, may be acceptably fast today. But the larger source of progress is due to better understanding of the problems that the computer is tasked with solving. More powerful computers make old methods faster and more practical, but they do not contribute to general intelligence, which requires something fundamentally different from mere capacity. Automobiles are not just faster horses.

Take, for example, weather forecasting. Weather forecasts have become amazingly more accurate over time. The accuracy of 5- to 7-day forecasts in 2015 was roughly equivalent to the accuracy of 1-day forecasts in 1965 (Stern & Davidson, 2015). Better computational capabilities have surely contributed to this increase in accuracy. But even more valuable was the benefit of better data, for example, more weather stations, and better dynamic models. Better computer capacity by itself would have merely sped up the process of making predictions. Better data and better models allowed those predictions to extend further into the future where they are more valuable.

Given these limitations on general intelligence, I am therefore somewhat puzzled by the concern of some philosophers and others that we are on the brink of creating a general artificial intelligence that will somehow displace humans in the world, like Skynet in the old Terminator movies.

> Let an ultraintelligent machine be defined as a machine that can far surpass all the intellectual activities of any man however clever. Since the design of machines is one of these intellectual activities, an ultraintelligent machine could design even better machines; there would then unquestionably be an "intelligence explosion," and the intelligence of man would be left far behind. Thus the first ultraintelligent machine is the last invention that man need ever make. (I. J. Good, 1965)

Good's hypothesis depends on the assertion that the ability to engineer new problem solutions is just like the ability to employ existing solutions, but these two kinds of problems are fundamentally different. Solving Einstein's famous equation is far different from coming up with the theory that it represents. Navigating the tree structure of chess or go is far different from having the idea of representing these games as trees. We know how to compute a solution to navigating a tree, but we do not yet know how to build a process that would have the insight to translate these games into a tree structure.

Solving well-structured, known problems is a process of selecting optimal alternatives from a menu or range of available choices. The choices may be discrete or they may be numerical, but all machine learning has this basic underlying structure. Special-purpose algorithms, like those used in existing examples of computational intelligence, have become increasingly capable of solving ever more complex problems of choices sort, but they still do not have the capability of inventing something from a new perspective. The evidence we have suggests that invention—for example, the design of new unforeseen structures, the formulation of new scientific paradigms, or the creation of new forms of representation—requires a different set of skills than optimization over a known space. We do not currently have any idea how to build a computer system that can come up with novel representations, but that ability is essential to achieve general intelligence.

Some of the fear concerning the potential for superintelligent machines to run amok comes from inconsistent thought experiments. For example, Nick Bostrom asks us to imagine an artificial intelligence machine that is given the goal of making as many paper clips as possible. It somehow becomes superintelligent, however, and rewrites its own capabilities to be even more intelligent about making paper clips. In following its maxim to make as many paper clips as possible, it converts everything it can to paper clips, in the process destroying the world.

I don't find this thought experiment very compelling, not because I think that paper clips are silly but because it both supposes that the machine is superintelligent and, at the same time, super focused on one thing—making paper clips. It is broadly intelligent; it has superior general intelligence, yet it is singularly stupid in being focused just on paper clips. If it were so smart, it could analyze its compulsion to produce paper clips. It is difficult to imagine that it could be superintelligent without this capability. It is difficult to imagine that anything could be so dramatically intelligent and so dramatically fixated on a single narrow task.

There are many other reasons to doubt the usefulness of Bostrom's thought experiment. We will treat this question in more depth in a later chapter. For now, it may be enough to note that a computer designed to make paper clips has no functions that would allow it to improve itself. Just as a go-playing computer is of no use to playing chess, it is difficult to see how a paper clip–making computer could be of any use to improving the computational intelligence of computers, including itself. They are different problems, and there is no bridging technology available in the current world or in Bostrom's thought experiment that would allow the computer to move from one to the other. It may learn to better navigate the space of paper-clip making, but that space does not include anything about improving computing. There currently is no method that would allow a chess-playing computer to claim boredom with the game and to then direct its efforts to reading poetry. Creating computers with that kind of capability will require approaches that are not being used, or perhaps not even being imagined today.

Superintelligence does not now exist and the current approaches to AI do not provide a path to get to it. Creating a superintelligent AI would require an approach that we have not yet conceived of. That is not to say that it is impossible, but it does say that we are not yet even heading in the right direction to achieve it. New approaches, invented by people, will be needed to achieve that goal.

This book is intended to provide an understanding of what is needed to achieve general intelligence. It is a road map for research, but not yet a report of the outcome of that research.

The current press coverage of artificial intelligence would have you believe that we are on the verge not only of general intelligence but of a runaway superintelligence that will first come for our jobs and then our babies.

Although it is true that computational intelligence is now capable of taking on a large number of tasks that have previously been performed by humans, it is also creating other new jobs that have never been available in the past. It has the potential to disrupt and change many jobs, but it will not destroy the economy in the process, just change it.

The prospects of an exponentially improving superintelligence that will destroy the world, as in Bostrom's paper-clip thought experiment, are zero as well. Machine learning may be speedier on faster processes, but ultimately, it depends on feedback from the world to know if something new actually works or not. Predicting the weather five days into the future requires that you wait five days to find out if it worked. Although old data may provide a good source for learning how to predict the weather, a forecast is only valuable if it tells us what the future weather will actually be. Faster computers cannot make the weather appear any faster, and so the speed at which a system can improve itself is limited by the speed at which data appear, not just the speed of its computations.

Even if we solve all of the problems associated with general intelligence learning, the rate at which it can evolve its capabilities is limited by the speed with which the world can provide feedback, and that is not affected by computer processing capacity. It has taken us 50,000 years to invent the current state of intelligence, there is no telling how long it would take to invent our way to general intelligence and then to superintelligence.

Resources

Aubert, M., Setiawan, P., Oktaviana, A. A., Brumm, A., Sulistyarto, P. H., Saptomo, E. W., . . . Brand, H. E. A. (2018). Paleolithic cave art in Borneo. *Nature, 564,* 254–257. https://www.nature.com/articles/s41586-018-0679-9.epdf

Bayern, A. M. P. von, Danel, S., Auersperg, A. M. I., Mioduszewska, B., & Kacelnik, A. (2018). Compound tool construction by New Caledonian crows. *Scientific Reports, 8,* 15676.

Beran, M. J., Rumbaugh, D. M., & Savage-Rumbaugh, E. S. (1998). Chimpanzee (*Pan troglodytes*) counting in a computerized testing paradigm. *The Psychological Record, 48*(1), 3–20. http://opensiuc.lib.siu.edu/tpr/vol48/iss1/1

Bostrom, N. (2014). *Superintelligence: Paths, dangers, strategies.* Oxford, UK: Oxford University Press.

Boysen, S. T., Berntson, G. G., Shreyer, T. A., & Hannan, M. (1995). Indicating acts during counting by a chimpanzee (*Pan troglodytes*). *Journal of Comparative Psychology, 109,* 47–51.

Calvin, W. (2004). *A brief history of the mind.* Oxford, UK: Oxford University Press.

Clark, A. (1998). *Magic words: How language augments human computation.* doi:10.1017/CBO9780511597909.011; http://www.nyu.edu/gsas/dept/philo/courses/concepts/magicwords.html

Gabora, L. (2007). Mind. In R. A. Bentley, H. D. G. Maschner, & C. Chippendale (Eds.), *Handbook of theories and methods in archaeology* (pp. 283–296). Walnut Creek, CA: Altamira Press.

Good, I. J. (1965). Speculations concerning the first ultraintelligent machine. In F. Alt & M. Rubinoff (Eds.), *Advances in computers* (Vol. 6, pp. 31–88). New York: Academic

Press. https://vtechworks.lib.vt.edu/bitstream/handle/10919/89424/TechReport05-3.pdf?sequence=1

Hofstadter, D. R. (1999) [1979], *Gödel, Escher, Bach: An Eternal Golden Braid.* New York: Basic Books.

Kaminski, J., Tempelmann, S., Call, J., & Tomasello, M. (2009). Domestic dogs comprehend communication with iconic signs. *Developmental Science, 12,* 831–837. https://www.eva.mpg.de/psycho/pdf/Publications_2009_PDF/Kaminski_Tempelmann_Call_Tomasello_2009.pdf

MacPherson, K., & Roberts, W. A. (2013). Can dogs count? *Learning and Motivation, 44,* 241–251.

Markham, J. A., & Greenough, W. T. (2004). Experience-driven brain plasticity: Beyond the synapse. *Neuron Glia Biology, 1,* 351–363. doi:10.1017/s1740925x05000219; https://www.ncbi.nlm.nih.gov/pmc/articles/PMC1550735

McCarthy, J., Minsky, M., Rochester, N., & Shannon, C. E. (1955). *A proposal for the Dartmouth Summer Research Project on Artificial Intelligence.* http://jmc.stanford.edu/articles/dartmouth/dartmouth.pdf

Neisser, U., Boodoo, G., Bouchard, T. J. J., Boykin, A. W., Brody, N., Ceci, S. J., . . . Urbina, S. (1996). Intelligence: Knowns and unknowns. *American Psychologist, 51,* 77–101. http://differentialclub.wdfiles.com/local--files/definitions-structure-and-measurement/Intelligence-Knowns-and-unknowns.pdf

Newell, A., Shaw, J. C., & Simon, H. A. (1958). Elements of a theory of human problem solving. *Psychological Review, 65,* 151–166.

Owano, N. (2013). Scotland lunar-calendar find sparks Stone Age rethink. Phys.org. https://phys.org/news/2013-07-scotland-lunar-calendar-stone-age-rethink.html

Pásztor, E. (2011). Prehistoric astronomers? Ancient knowledge created by modern myth. *Journal of Cosmology, 14.* http://journalofcosmology.com/Consciousness159.html

Pearl, J., & Hartnett, K. (2018). To build truly intelligent machines, teach them cause and effect. *Quanta Magazine.* https://www.quantamagazine.org/to-build-truly-intelligent-machines-teach-them-cause-and-effect-20180515

Pearl, J., & Mackenzie, D. (2018). *The book of why: The new science of cause and effect.* New York: Basic Books.

Silver, D., Huang, A., Maddison, C. J., Guez, A., Sifre, L., van den Driessche, G., . . . Hassabis, D. (2016). Mastering the game of go with deep neural networks and tree search. *Nature, 529,* 484–489. http://airesearch.com/wp-content/uploads/2016/01/deepmind-mastering-go.pdf

Simon, H. A., & Newell, A. (1971). Human problem solving: The state of the theory in 1970. *American Psychologist, 26,* 145–159. https://pdfs.semanticscholar.org/18ce/82b07ac84aaf30b502c93076cec2accbfcaa.pdf

Smithsonian National Museum of Natural History. (2016). Human characteristics: Brains: Bigger brains: Complex brains for a complex world. http://humanorigins.si.edu/human-characteristics/brains

Stern, H., & Davidson, N. E. (2015). Trends in the skill of weather prediction at lead times of 1–14 days. *Quarterly Journal of the Royal Meteorological Society, Part A, 141,* 2726–2736.

Sternberg, R. J., & Detterman, D. K. (Eds.). (1986). *What is intelligence?* Norwood, NJ: Ablex.

2 Human Intelligence

In this chapter we consider just what it means for a human to be intelligent. Computers do not have to solve the same problems in precisely the same way, but it is still necessary to understand just what problems human intelligence does solve. General intelligence must still solve the same range of problems that a human can solve, so understanding that range is a critical step in creating general intelligence.

Human intelligence is our best known example of an intelligent system. In the early days of computational intelligence, following the 1956 Dartmouth workshop, the goal was to describe every aspect of human intelligence with enough precision that it could be simulated on a machine. Since that time, many working in the field have found that practical applications of computational intelligence do not need to duplicate how people solve problems, but rather these workers have found ways to reduce the complexity of an intelligence task to something that can be accomplished by a computer. General intelligence, on the other hand, does not seem to be solvable in the same reductionist way. General intelligence may actually gain from a deeper understanding of the best example we have of general intelligence—us.

As discussed in the introduction, conceptions of human intelligence focus on tasks that we associate with higher cognitive functioning—the kind of tasks that the people whom we admire for their superior intelligence perform that we cannot. The ability to do work in the field of theoretical physics, the ability to compose great music, and the ability to play chess are among these. These characteristics involve tasks that have been invented by people over time, and they are tasks that usually require formal education.

These intellectual processes have long been thought to be the basis not only of human intelligence but, essentially, of human thought. George Boole (1854) titled his famous book on logic *The Laws of Thought.* In this he echoed Aristotle's *Organon,* in which Aristotle described three fundamental laws (identity, non-contradiction, and excluded middle), which, he argued, were the essential basis of logic and thought. John Stuart Mill (1836/1967) described a view that was later called economic man, who made rational decisions to support his quest for wealth.

We have always known that people do not always behave in the systematic ways suggested by these views of logical thought, but these deviations were attributed to intrusion by emotions in the thought process. As we will see, they are more properly viewed not as glitches or bugs in human thought but as essential features that enable human intelligence.

There are currently no widely accepted definitions of just what it means for a human to be intelligent. Shane Legg and Marcus Hutter (2007) list 70 definitions of intelligence. Most of these definitions emphasize higher mental function, such as the ability to think rationally, to reason, to plan, and to solve problems. They include the ability to think abstractly, to learn quickly, and to comprehend complexity. They are characteristic of the kind of skill that allows one to succeed in Western society. It is no coincidence, therefore, that these are the kind of skills also examined by intelligence tests.

Intelligence Testing

The work on human intelligence has largely been focused on understanding individual differences among people. Alfred Binet, for example, was asked by the Paris school system at the start of the twentieth century to find a way of identifying those students who would need more help in getting an effective education. To answer this question, he and his colleague, Theodore Simon, created a test that sought to evaluate characteristics that they believed were indicators of the constituents of intelligence, including measures of language skills, memory, reasoning, and the abilities to follow commands and to learn new associations. They were looking for a measure that was independent of the amount of schooling a child had already received, so they avoided tests of specific facts and other kinds of explicit knowledge. The kinds of characteristics that they tested were soon expanded into other measures of intelligence for many other purposes.

Binet recognized the limitations of his test. He also recognized that the single number score for a test could not do justice to a student's actual intelligence, but it was a reasonably good predictor of how well students would do in school. In contrast, an English psychologist, Charles Spearman (1904), proposed that intelligence is actually a unitary quality that could be measured by a proper test of general intelligence and could be represented by a single number.

Most intelligence tests include assessments of a number of nominally individual skills. Spearman noted that people who score well on one of the subtasks typically perform well on others, and those who perform poorly on some tend to perform poorly on others (see chapter 1). Spearman invented some new statistics to evaluate this correlation. He used this new method, called "factor analysis," to statistically divide the test takers' performance into two kinds of components or factors. In Spearman's view, the student's performance on each specific subtask in the intelligence test is due to some combination of a specific "intelligence" associated with that subtask and a general "intelligence," "g," that contributes to performance across many subtasks. The correlation between subtasks, he thought, is caused by the fact that they share the same general intelligence factor. If someone has more of it, that person will tend to score well on most of the tests, and if someone has less, that person will score poorly on most of the tests.

Spearman argued that the general factor was the result of some biological characteristic of human brains or minds, something akin to mental power. Some psychologists have attributed general intelligence to brain size, or mental speed, measured by the speed with which simple decisions can be made. Others have attributed it to such factors as memory capacity or visual acuity.

From a statistical point of view, something must be shared among the correlated tasks, but that something need not actually be intelligence. Anxiety, calm, experience with test-taking methods, attention or motivation, among others, could be the common factor.

On the other hand, the correlation may reside in the similarities between the tests themselves. Tests that appear to be different may still share overlapping skills. For example, two of the tests used to assess intelligence are the number sequence task and the progressive matrices task. In a number sequence task, the test taker has to guess the number that follows a

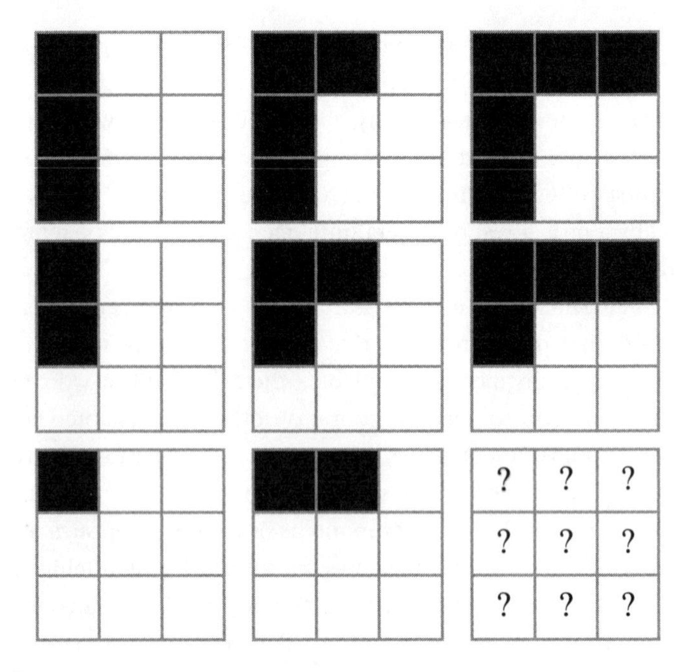

Figure 1

A simple example of a progressive matrix task used to assess intelligence. What pattern should be drawn in the ninth box that would be consistent with the previous squares in the row and in the column?

presented set of numbers (for example, what number would follow the sequence 2, 4, 6, 8?). In a progressive matrices task (see figure 1), the student is shown a matrix of designs exhibiting a certain pattern and must draw or choose the final design in that sequence. Both tasks require the student to induce the rule for the respective pattern and apply that rule. They both, in other words, tap some overlapping set of skills, and this overlap could be the cause of the correlation.

The jury is still out on whether there is such a *thing* in humans as general intelligence, at least as measured by intelligence tests. Computer scientists and psychologists have both been searching for it, but it has so far proven to be elusive.

Intelligence, as measured by intelligence tests, has been found to correlate with many intellectual capabilities, but not always the ones you might expect. It seems, for example, to have a weak relationship, if any, to complex problem-solving ability (Wenke, Frensch, & Funke, 2005).

Problem Solving

The ability to solve problems is a common feature among definitions of intelligence. Fortunately, this capability has also been well studied by psychologists and may provide an alternative means to get at the nature of intelligence.

Well-Formed Problems

In order for testers to be able to score intelligence tests, the tests must consist of specific questions that have specific answers. Real-world problems, on the other hand, often involve a large number of potential variables in complex relations. The goals of real-world problems may be unclear, and a substantial part of solving them is just finding the right goals. Studies of human problem solving involve well-formed problems because they are easy to administer, easy to score, and relatively easy to understand.

These laboratory tasks involve well-understood problems, and their outcomes are easy to evaluate. Games like chess, and now go, are complex, but they are very well-defined by their rules and by the position of the pieces during the game. There may be a lot of potential moves, but all of the valid moves are easy to identify.

Although there are laboratory studies of how people play chess, many psychological studies of problem solving have focused on simpler well-formed problems to be able to examine the entire problem-solving process in a reasonable amount of time. Three of these are the 8-tile problem, the Towers of Hanoi problem, and the hobbits and orcs problem (all three problems will be described shortly). These are simple enough to be solved in a brief laboratory session; the state of the problem is easy to describe without uncertainty. Finally, they do not rely on any particular knowledge to be able to solve them.

The 8-tile problem consists of a square frame containing eight tiles, numbered 1 through 8, and one empty spot. The digits are originally in some random order, and the solver's task is to arrange them in numerical order. The initial order is the "starting state," and the correct numerical order is the "goal state." Each step in solving the problem consists of moving one of the tiles into the empty slot. Only one tile can be moved at a time, and only a tile adjacent to the empty slot can be moved. Given a starting position, we could exhaustively list the succession of possible moves. We could even draw a diagram of those possible moves. Each specific arrangement of the tiles is a "state" and the set of all possible arrangements is the "state space" for the problem. As in chess, the problem can be represented as a tree (see chapter 1), where each choice is a branch of the tree.

We solve the problem by successfully moving through this state space from the starting position through some sequence of selected states (by moving a tile) and finally reaching the goal state. We could choose a path through the state space by selecting the move at each point that gets us closer to the goal state.

Here is an example of one starting configuration. The empty tile is in the middle row and middle column:

1	4	3
7		6
5	8	2

From this configuration, there are four possible moves. We could move either the 4-tile, the 6-tile, the 7-tile, or the 8-tile into the blank space because these numbers are adjacent to the empty space. If the 4-tile is chosen, then the empty space will be in the center of the top row, as shown by the next configuration:

1		3
7	4	6
5	8	2

Then, on the next step, either the 1- or 3- or 4-tile could be moved, and so on.

The second commonly studied problem is the so-called Towers of Hanoi problem. See figure 2.

The puzzle was first described by Eduardo Lucas in 1883. In Lucas's version, the towers were supposed to be in an Indian temple dedicated to Brahma. In the more commonly known version, described by Sam Loyd

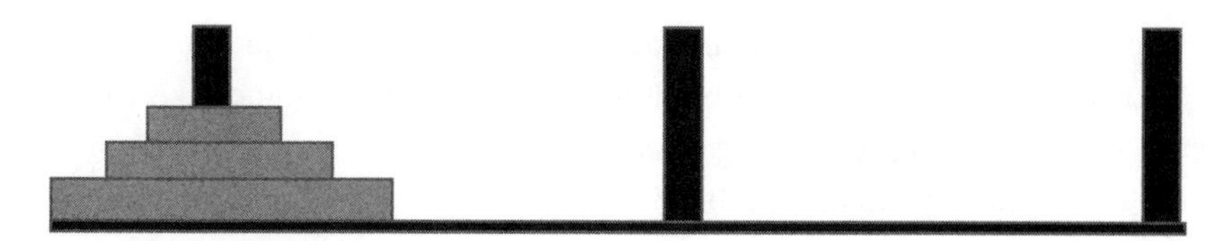

Figure 2
The three disk version of the Towers of Hanoi problem. The goal is to move the three disks from the first spindle to the last spindle following the rules of the task.

(1914), it was described as a problem being solved by monks in a fictitious temple in Hanoi, Vietnam. Supposedly, in the temple, the monks have to move a stack of 365 disks from one spindle to another. In the laboratory version, only three disks are typically used.

The laboratory version consists of three spindles and three disks of varying sizes. The starting state has the three disks stacked onto spindle 1 with the largest disk on the bottom and the smallest disk on the top. The puzzle solver's job is to move the disks from the first spindle to the third one, while obeying certain rules. Only one disk can be moved at a time, only one disk can be off of a spindle at a time, and a larger disk can never be placed on top of a smaller one (see, for example, Anzai & Simon, 1979, who studied solving a five-disk, three-spindle version of this problem).

With three disks and three spindles, there are only a few possible states. Initially, all three disks are on the first spindle. With three disks, the problem can be solved in a minimum of seven moves:

1. Move the smallest disk to the third spindle.
2. Move the medium disk to the middle spindle.
3. Move the small disk to the middle spindle.
4. Move the large disk to the third spindle.
5. Move the small disk to the first spindle.
6. Move the medium disk to the third spindle.
7. Move the small disk to the third spindle, and we are done.

As with the 8-tile problem, the number of states with three disks can be listed out explicitly. The problem is small enough to be solved in a short laboratory session. As the number of disks increases, though, the minimum number of moves needed to solve it grows exponentially. With 64 disks, and a move every second, it would take 585 billion years to solve.

The number of moves essentially doubles with each additional disk. Even though solving the puzzle with a large number of disks would take a very long time, the rules for solving it are easy to describe.

In the hobbits and orcs problem, three hobbits and three orcs arrive at a riverbank, and they all wish to cross to the other side (see Jeffries, Polson, Razran, & Atwood, 1977). There is a boat, but it can hold only two creatures at a time (two hobbits, two orcs, or one of each). If the orcs on one side of the river outnumber the hobbits, they will eat the hobbits, so you must be sure that there are never more orcs than hobbits on either side of the river. Other than the orcs' uncontrollable appetite for hobbits, all six of the creatures arriving at the river can otherwise be trusted. How can you get the six creatures across without losing any hobbits?

Here is a solution to this problem. "H" represents a hobbit. "O" represents an orc. The arrangement of hobbits and orcs on each side of the river constitutes the state of the problem, and the boat represents the transitions between states. See table 1.

These three simple problems, like the more complex ones such as go, chess, or checkers, are called "path problems." They can be described by a set of states and a set of actions (called "operators") for moving from one

Table 1

Description	Left Bank	Right Bank
All six arrive at the river	OOO HHH	
Send 2 orcs across	O HHH	OO
1 orc returns with the boat	OO HHH	O
Send 2 orcs across	HHH	OOO
1 orc returns with the boat	HHH O	OO
Send 2 hobbits across	O H	OO HH
1 hobbit and 1 orc return with the boat	OO HH	O H
Send 2 hobbits across	OO	O HHH
1 orc returns with the boat	OOO	HHH
Send 2 orcs across	O	OO HHH
1 orc returns with the boat	OO	O HHH
Send 2 orcs across		OOO HHH
Problem solved		Goal state

state to the next. Solving the problem amounts to finding a path from one state to the next and ultimately to the goal of the problem. These are the kinds of problems that Allen Newell and Herbert Simon (1972) used to derive their computer simulation of human problem solving, which they called the General Problem Solver.

The system starts in the initial or starting state and has solved the problem when it reaches the goal state. For example, in the hobbits and orcs problem, the state is the number of hobbits and the number of orcs on each side of the river and the position of the boat. Applying an "operator" leads to a change in state. In the hobbits and orcs problem, the operator is to send the boat from one side of the river to the other with some beings in it. Solving the problem means finding a sequence of states that lead from the starting state to the goal state. We can say that problem solving in this framework consists of a "search" of the state space (the set of all possible states and operators) to find a path through it.

When you apply the right operators to the states in the right order, you have solved the problem. The entire problem-solving process can be reduced to finding this correct path through the state space.

Formal Problems

The path problems that we have been talking about can also be called "formal problems." "Formal," in this context, means that it is the form of the problem, rather than the specific content or the physical properties of the problem, that determines how it can be solved. There are no real hobbits and no real orcs, for example, that we can send across the river, but we solve the problem with symbols. Those same symbols could stand for missionaries and cannibals instead of hobbits and orcs, and the problem would be formally identical and would be solved in exactly the same way.

Path problems typically involve specific rules, and the states, goal, and operators are defined for the solver. "All" the solver has to do is to find the path through the states that leads to the goal. It is usually unambiguous as to what the current state of the problem is. For example, how many hobbits and how many orcs are on each side of the river and where the boat is are all clearly known. There may be a large number of potential next steps (operators) that could be applied, but there is no ambiguity as to what actions are available.

Chess can also be described as a path problem; soccer, another competitive game, cannot. Although soccer has a clear goal (to outscore your opponent), an unambiguous way of assessing goal achievement, and rules for the kinds of things you can do, its state space is vastly more complicated because every other player could be anywhere on the pitch (field). Players and the ball are not restricted to a list of possible places they could be in the same way that chess pieces are restricted to one of 64 specific positions on the board. The ball does not always go where intended, meaning that there is uncertainty when an operator, that is, a kick, is applied. Teammates may not be where a passing player thinks they are, so there is uncertainty about the state of the game, and so on.

With path problems you can usually tell whether or not you are making progress toward your goal. But there are other problems where it is not so easy to know whether you have made progress. For example, when determining how to reduce poverty, there may be no clear method that unambiguously tells the policy maker whether a given plan is working or not.

Nonpath problems cannot typically be described as a step-by-step process. Instead, they often require some reorganization of thinking. Before that reorganization, reaching the goal may be impossible; afterward, it may be easy and obvious. Would, for example, just giving everyone some money solve poverty? I don't know.

The fundamental issue with Newell and Simon's approach to problem solving, and with much of computational intelligence investigations since then, is that they treat intelligence as a formal path problem. They assume, for example, that logic is a model for human thinking, that states are unambiguously known, and that operators always produce the expected effect.

Standard Boolean logic is a formal system; it can be described as a set of axioms and rules of inference. Applying an inference rule is an expression that results in a new state (a new arrangement or expression of symbols) in the same way that choosing a move in chess moves the system from one state to the next. Only certain inferences are valid/legal, and so only certain states are reachable. The correctness of an expression in a formal system, such as logic, depends on its form, not what the expression is about. Correctly formed expressions are necessarily correct. The content of those statements is not relevant.

If the premises of a syllogism (a Boolean logic expression) are true and the syllogism is of the right form, then the conclusion must also be true:

Premise: Bossy is a cow.

Premise: All cows are mortal.

Conclusion: Therefore Bossy is mortal.

Newell and Simon's General Problem Solver was a formal system in that it consisted of a set of basic tokens (axioms) and rules to manipulate them to make inferences. Games like checkers, chess, or go are formal systems because they consist of the basic pieces (the board and the playing pieces) and rules by which they can be manipulated. The pieces may have some meaning (for example, the knight and the bishop of chess), but one could effectively play chess without knowing their meaning, or even without any physical pieces at all.

The board and the positions of the chess pieces can be represented symbolically. For example, on one notation, each square on the chessboard is represented by a letter, indicating the square's column, and a number, indicating the square's row, similar to how we denote the cells in a spreadsheet. Each piece is represented by an uppercase letter, for example, Q for queen, R for rook (castle). A move is expressed by the symbol for the piece and the coordinate to which it is moved. The move Be5 means to move a bishop to the square e5. The whole game can be conducted using this symbolic notation or some other notation without ever touching physical pieces or a physical board.

Although formal reasoning is very important to intelligence, it is not all there is. In the next chapter, we will take up this question from a computational perspective. From a human cognition point of view, however, the evidence is clear that people do not inherently think logically. Logical thinking takes special effort.

Intelligence and formal reasoning imply rational decision-making. They imply that the reasoner will choose operators that advance it toward the goal. In general, a rational decision is one that is based on objective facts and that maximizes a desired benefit. Unless we are willing to just make up willy-nilly goals that fit whatever a person does, human decision-making often fails to be rational. Some people smoke, even though they know that there are health risks involved. We can imagine that there must be some goal that is rationally furthered by smoking, but that is circular reasoning. It makes up the goal to match the action and then tries to explain the action by this made-up goal. There may be some goal that is rationally furthered by jumping out of perfectly good airplanes or by leaping on a grenade to save one's comrades. That last one may be heroic, but it is not in the personal interest of the hero to do it—it appears to be irrational.

Rational decisions are based on solid evidence and statistics. Rational decisions are often the more intelligent choice. People who make better, more rational decisions are usually perceived as being more intelligent than those who do not. One of the roles that logic plays, for example, is to help people reason systematically about the choices that they make. If the form is right, then the right decision should be consistently reached if people were rational decision makers. But they are not, at least not always.

For example, Amos Tversky and Daniel Kahneman found a number of situations in which people fail to make rational decisions. For instance, they found that people make different decisions under formally identical situations depending on how that situation is described. An example of this is that when graduate students were told of a penalty for registering for a conference after a particular date, 93% of them registered early, that is, before that date. When offered an identical early registration discount (that is, one with the same price difference before versus after the date), only 67% of them registered before that date. The two situations are identical, with the same benefit for registering early. The only difference was the label given to the action (penalty versus discount), but this label made a substantial difference. The students sought to avoid a loss described as a penalty but did not go out of their way for a gain.

Historically, this difference would have been interpreted as evidence that emotion intruded on the decision-making process and led the students to make an emotional rather than a logical decision. There is another possibility, however, that suggests that this deviation from rational decision-making was not a failure, but evidence for other processes that may play a role in intelligence. In fact, a formal system cannot be sufficient, even for logical reasoning.

A formal system depends solely on its internal structure, but intelligence requires interaction with a world, a world that includes uncertainty. A formal system starts with a set of basic premises, assumptions, or axioms. If the axioms are true, and the statements are of the right form, then the conclusions must also be true. The formal system assumes that the axioms are true, but there is no guarantee that this assumption is correct. The formal system depends on the truth of the axioms, but by itself, it cannot establish their truth.

In logic, the axioms are typically called "premises." The premises could be wrong. For example, in the cow syllogism, we could assume that Bossy is a cow. We could further assume that all cows are mortal. Using the rules of the system, we could then infer that therefore Bossy is mortal. So far, so good, but how do we know that Bossy is actually a cow? That assumption could be wrong and there is no formal method to prove that it is true. If the axioms are not true, then any conclusions derived from those faulty axioms may also be faulty. If Bossy only looked like a cow but was actually an advanced robot, she might not, in fact, be mortal.

We might do tests to show that Bossy is a cow. But no matter how many tests we did, and no matter how many she passed, there is still a chance that we could be wrong, that the very next test we ran would indicate that she is a robot and not a cow.

We cannot prove that an axiom or premise is actually true. Deductions can be proved from the premises, but the premises cannot. We cannot infallibly move from specific observations to general truths. That inference must transcend logic. It depends critically on real-world facts, and there is no formal system that can prove that those facts are correct.

Starting in the late 1920s, a group of philosophers tried to create an approach to science that was strictly logical. In their view, scientists were misled when Newtonian mechanics was "replaced" by quantum mechanics. The basic principles of physics were not as Newton had described them. The logical positivists, as this group was known, tried to reduce science to just observation statements and logical deductions from those observations. If they could eliminate the sloppy language that was inherent in scientific theories, they argued, science would never be deceived again.

Observation statements (like "The temperature of the mixture increased by 2 degrees"), they thought, could be infallible as long as they were made with a healthy mind, that is, they ruled out hallucinations and the like as valid observation statements.

Without getting too far into the philosophical details, the approach of logical positivism failed. No purely logical system could produce science. Observations could be mistaken. Not every scientific statement could be immediately verified. As Kurt Gödel showed, not even mathematics, the most systematic and logical approach to knowledge that there is, could survive as a complete system based solely on observation statements and logical deductions from them. Thomas Kuhn and later Imre Lakatos countered the logical positivists with a more psychological approach to scientific thinking.

Therefore, if the two examples that were arguably the most typical of human intelligence could not survive based on pure logic, it is extremely unlikely that similar processes could be the sole cause of human intelligence. Human intelligence has to go beyond mere observation and deductions from those observations.

Establishing the truth of a premise requires an inference. Inferences are always subject to uncertainty. We might think that we are playing a game of chess, but if, in fact, it only looked like chess, then the formal properties of the game might be different and success of the formal system would fly out the window.

Much of the science of computer science derives from treating computer algorithms as formal systems that can be proven to be true. An algorithm does not care what the computations represent, only that it is in the right form, and, if it is in the right form, it can be proved to be correct. The meaning of the variables in an algorithm does not affect the validity of the process. Two plus two equals four whether it is two ducks, two trucks, or two bucks. Algorithms do not care what they are reasoning about, but people often do.

Unlike formal systems, human intelligence often depends critically on the content of what we are thinking about. Humans are capable of believing things that are not true. Human language can express sentences that are neither true nor false, such as "This sentence is false." Humans interact with an uncertain world.

People have to go to school to learn logic, and many people find it difficult. If logic were the basis of human thought, then it would come "naturally," like walking. People who are educated to take advantage of formal systems are often able to accomplish tasks that they would not be able to do without such tools. On the other hand, simpler, more intuitive processes can often succeed where complicated formal systems would either take too long or be unduly affected by irrelevant information.

As discussed in chapter 1, people employ heuristics to guide much of their thought. A heuristic is a practical method that generally works, but, unlike an algorithm, is not guaranteed to produce the correct result. Typically, for example, the taller child is likely to be the older child, but this heuristic can sometimes be wrong. One of the values of heuristics is that they allow people to reach conclusions that may not be fully justifiable but still may be valuable. The conclusion may not be provable, but it may take only a small amount of effort to reach it, and still be accurate enough for practical purposes. Because heuristics sometimes fail, they may also lead to false conclusions and prejudices that can sometimes interfere with intelligent action. They can have both value and cost and still contribute positively.

One heuristic that people use is called the "availability heuristic." People base their judgment on the examples that they can most easily bring to mind. Items that can be recalled most easily are treated as if they were the most representative examples and, therefore, the most important examples for making decisions.

The availability heuristic depends on unwarranted assumptions, but practically speaking, it can often be an effective way of dealing with real-world situations. Often the easiest to remember items are, in fact, the most relevant to the judgment. For example, if judging whether Chicago or Boston is the larger city, a full analysis might give a good answer, but availability might also provide an answer.

Under certain circumstances, the consequences of using the availability heuristic can sometimes conflict with a well-reasoned analysis, but under other circumstances, its use may be at least as accurate as a formal process. Unlike a detailed analysis, heuristic answers are often much faster and require much less effort than an exhaustive analysis.

If you were using availability to choose the larger city, you would decide that Chicago is the larger city if facts about it are more available than facts about Boston. If it is easier to call to mind facts about one city than another, the more fact-related city is likely to be the bigger one.

We cannot know directly how available memories are of these two cities for any specific person, but we can use another heuristic to estimate availability. We can, for example, look at the number of mentions each of these cities has in Google. The thinking is that if a city is mentioned more often in Google, then it is likely to be easier to think of the facts that are mentioned. This too, is a heuristic.

A Google search for "Chicago" at the end of 2019 claimed about 3 billion hits and a similar search for "Boston" claimed about 1.9 billion. Also according to Google, the population of Chicago is listed as 2.7 million, and that of Boston as 685,094. If Google mentions are a good estimate of availability then it would seem sensible to predict that availability is a cheap and quick way to estimate population size.

The same heuristic can be used to correctly identify the better basketball team. In 2017, the Cleveland Cavaliers were in the National Basketball Association (NBA) final playoffs, but the Atlanta Hawks were also-rans. Again, using Google search results as an estimate of availability, the Cavaliers received 12.7 million hits and Hawks received 3.6 million hits in 2017. So, again, these heuristics work.

Heuristics such as availability apparently evolved because these low-cost estimates are often effective. Using them may not be a failure of rational thinking, but a success of natural cognition. They may provide an effective and efficient adjunct to the invented intelligence so commonly taught.

Heuristics like these may be critical to creating broadly effective computational intelligence, but computer science has focused largely on well-structured formal problems, like chess playing and theorem proving. It has run into challenges when dealing with less structured problems like driving or facial recognition. Recent success in these less formal problems has come from the recognition that heuristic tools can be used effectively even if they entail greater levels of uncertainty. Neural networks, for example, are less formal than expert systems. They exploit continuous nonsymbolic representations that only approximate the state of the world; they do not symbolize it. They sacrifice provability for improved accuracy under more naturalistic circumstances.

Insight Problems

As mentioned earlier, the emphasis of intelligence testing and computational approaches to intelligence has been on well-structured and formal problems. These problems may be complex, but they are easy to understand and easy to evaluate. But the focus on these well-structured problems may be like an attempt to look for your lost keys where the light is brightest. There are other problems that are typical of intelligence that do not fit into this well-structured framework.

A major critical group of nonformal problems that people face are the so-called insight problems. Insight problems generally cannot be solved by a step-by-step procedure, like an algorithm, or if they can, the process is extremely tedious. Instead, insight problems are characterized by a kind of restructuring of the solver's approach to the problem. In path problems, the solver is given a representation, which includes a starting state, a goal state, and a set of tools or operators that can be applied to move through the representation. In insight problems, the solver is given none of these. Solving insight problems depends on discovering the representation appropriate to the problem, and once that representation is discovered, the solution is typically easy and rapid.

A typical insight problem is the one that supposedly led Archimedes to run naked through the streets of Syracuse when he solved it. As the story goes, Hiero II (270 to 215 BC), the king of Syracuse, suspected that a votive crown that he had commissioned to be placed on the head of a temple statue did not contain all of the gold it was supposed to. Archimedes was tasked with determining whether Hiero had been cheated. He knew that silver was less dense than gold, so if he could measure the volume of the crown along with its weight, he could determine whether it was pure gold or a mixture. The crown shape, however, was irregular, and Archimedes found it difficult to measure its volume accurately using conventional methods.

According to Vitruvius, who wrote about the episode many years later, Archimedes realized, during a trip to the Roman baths, that the more his body sank into the water, the more water was displaced. He used this insight to recognize that he could use the volume of water displaced as a measure of the volume of the crown. Once he achieved that insight, finding out that the crown had, in fact, been adulterated was easy.

The actual method that Archimedes used was probably more complicated than this, but this story illustrates the general outline of insight problems. The irregular shape of the crown made measurement of its volume impossibly difficult by conventional methods. Once Archimedes recognized that the density of the crown could be measured using other methods, the actual solution was easy.

With path problems, the solver can usually assess how close the current state of the system is to the goal state. Most machine learning algorithms depend on this assessment. With insight problems, it is often difficult to determine whether any progress at all has been made until the problem is essentially solved. Insight problems are often associated with a subjective feeling of "Aha," as the solution is discovered.

Another example of an insight problem is the socks problem. You are told that there are individual brown socks and black socks in a drawer in the ratio of five black socks for every four brown socks. How many socks do you have to pull out of the drawer to be certain to have at least one pair of either color? Drawing two socks is obviously not enough because they could be of different colors.

Many (educated) people approach this problem as a sampling question. They try to reason from the ratio of black to brown socks how big a sample they would need to be sure to get a complete pair. In reality, however, the ratio of sock colors is a distraction. No matter what the ratio, the correct answer is that you need to draw three socks to be sure to have a matched pair. Here's why:

With two colors, a draw of three socks is guaranteed to give you one of the following outcomes:

Black, black, black—pair of black socks

Black, black, brown—pair of black socks

Black, brown, brown—pair of brown socks

Brown, brown, brown—pair of brown socks

The ratio of black to brown socks can affect the relative likelihood of each of these four outcomes, but only these four are possible if three socks are selected. The selection does not even have to be random. Once we have the insight that there are only four possible outcomes, the problem's solution is easy.

Insight problems are typically posed in such a way that there are multiple ways that they could be represented. Archimedes was stymied as long as he thought about measuring the volume of the crown with a ruler or similar device. People solving the socks problem were stymied as long as they thought of the problem as one requiring the estimate of a probability. How you think about a problem, that is, how you represent what the problem is, can be critical to solving it.

Interesting insight problems typically require the use of a relatively uncommon representation. The socks problem is interesting because, for most people, the problem is most likely to evoke a representation centered on the ratio of 5:4, but this is a red herring. The main barrier to solving insight problems like this is to abandon the default representation and adopt a more productive one. Once the alternative representation is identified, the rest of the problem-solving process may be very rapid. Laboratory versions of insight problems generally do not require any specific deep technical knowledge. Most of them can be solved by gaining one or two insights that change the nature of how the solver thinks about the problem.

Most of the problems given to computers for solution are well-structured path problems. The designer of the program provides the problem, its representation, and the operations that can move the computer toward its goal. It may be difficult to find a path to solution, using the representations, operators, and paths, because of the large number of possible states involved, but it is still a process of searching for and following a path. Insight problems, on the other hand, generally do not have a clear path. Computational intelligence research has not given serious attention to problems like these, but they are a clearly among the kinds of problems that an intelligent agent would have to address.

Here are a few more insight problems. The mutilated checkerboard was first described by Max Black in 1946. A regular checkerboard has 32 black squares and 32 red squares. If we had 32 dominoes, each the size of two squares, it would be obvious that we could cover the checkerboard with those 32 dominoes, for example, using 8 rows of 4 dominoes each. If we cut off the red square at the upper left corner of the checkerboard and the red square in the lower right corner of the checkerboard, could we now cover the mutilated checkerboard with 31 dominoes?

Another insight problem, the Königsberg bridges problem, is shown in figure 3. The city of Königsberg (now called Kaliningrad, Russia) was built on both sides of the Pregel River. Seven bridges connected two islands and the two sides of the river. Can you walk through the city, crossing the seven bridges each exactly once? In the map in figure 3, the bridges are marked in gray.

Here is a sequence of four numbers: 8, 5, 4, 9. Predict the next number in this sequence.

The two-strings problem was studied by Maier (1931). You are in a room with two strings hanging from the ceiling. Your task is to tie them together. In the room with you and the strings are a table, a wrench, a screwdriver, and a lighter. The strings are far enough apart that you cannot reach them both at the same time. How can these strings be tied together?

For the mutilated checkerboard problem we find that 8 rows of 4 dominoes will not work because two of the rows are short half a domino, but

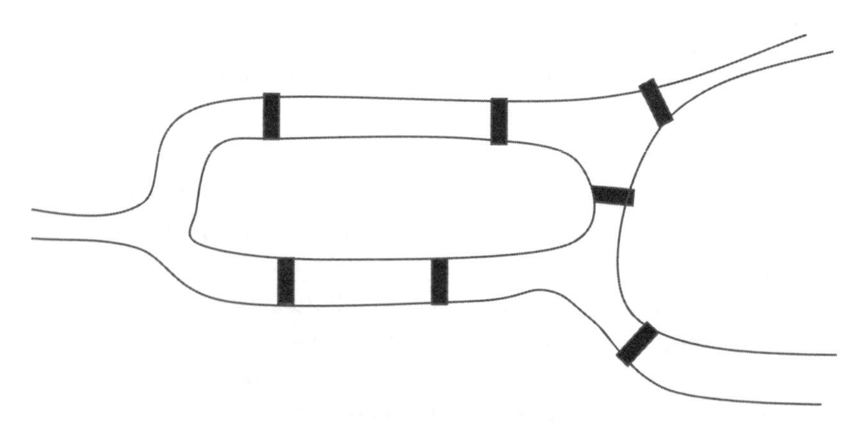

Figure 3
A sketch of the bridges connecting the land areas in Königsberg. Can you cross all seven bridges exactly once?

perhaps there is some arrangement of dominos that might work. You could try to lay out real or imaginary dominoes on the mutilated board, but when a particular pattern did not work, you would not know whether it was that pattern that was no good or whether there is no pattern that would work. Representing the problem in terms of dominoes and layouts makes solving the problem difficult at best. In theory, a computer could use this rearrangement method to try to determine whether the board can be covered by 31 dominoes, but it requires testing all possible arrangements. In the absence of insight, we have only brute force. There are no approximate solutions that can be used to help us search the tree of possible arrangements. We just have to try them.

Before we go back to the mutilated checkerboard problem, consider this one. There are 32 men and 32 women at a dance. Only heterosexual couples dance. Can everyone at the party dance at the same time? Now two of the women leave the party. Can we still form 31 heterosexual couples?

In the original checkerboard, each domino covered exactly one red square and one black square. Each heterosexual dance couple must contain exactly one man (black square) and one woman (red square). In the mutilated checkerboard, there are 32 black squares but only 30 red squares. Representing the problem this way reveals that it is impossible to cover a mutilated checkerboard exactly with 31 dominos even though there are exactly 62 squares. The mutilated checkerboard problem is formally identical to the heterosexual dance problem. People tend to find the dance problem relatively easy but find the checkerboard problem relatively difficult.

The mutilated checkerboard problem can be solved using a brute-force solution where every layout of the dominoes is tried. Trying a few thousand potential layouts may be practical with an 8 × 8 board but may not be practical with a much larger analogous board. There are 6,728 ways to arrange dominoes on a regular 8 × 8 checkerboard. But if we increase the number of squares to form a 12 × 12 "checkerboard," the number of possible domino arrangements grows to 53,060,477,521,960,000. With the insight that a domino must cover exactly one red and one black square, on the other hand, we can instantly solve the problem no matter how many squares are on the board.

An expert might recognize the mutilated checkerboard and the dance party problem as examples of a parity problem and solve both of them even more quickly. The dance party problem is much easier to solve because the useful representation is much more obvious, meaning that people are likely to come up with it quickly. Solving the dance problem can help solve the checkerboard problem if you can see the relationship between the two problems. Current approaches to computational intelligence generally cannot take advantage of this analogy. To be fair, many people fail to see the connection as well (Gick & McGarry, 1992).

The Königsberg bridges problem is also similar. Königsberg is divided into four regions. Each bridge connects exactly two regions. Except at the start or the end of the walk, every time one enters a region by a bridge, one must leave the region by a bridge. The number of times one enters must equal the number of times one leaves it, so the number of bridges touching a land mass must be an even number to cross them all exactly once because half of them will be used to enter a region and half will be used to leave it. The only possible exceptions are the regions where you start your walk and where you end it. Only a city with exactly none or exactly two regions with an odd number of bridges (one where you start and one where you finish) can be walked without repetition. In Königsberg, each region is served by an odd number of bridges, so there is no way that one can walk the seven bridges exactly once.

The checkerboard, dance, and bridges problems are related. They can all be represented as graphs (nodes connected by arcs). For our purposes, these three problems illustrate two things. How you represent the problem can profoundly affect the ease of solving it, and transfer from one problem to another is facilitated if you can find the analogy between the two of them.

If you had trouble with the digit sequence problem mentioned earlier, try writing out the names of digits in English:

Eight five four nine

The correct answer is 1, 7, 6.

The full sequence is:

Eight five four nine one seven six three two zero.

They are listed in alphabetical order of their English names. The usual representation of the series as digits ordered numerically must be replaced by a representation in which the English names are ordered alphabetically.

The string problem can be solved by using one of the tools as a weight at the end of one of the strings so that you can swing it and catch it while holding the other string. The insight is the recognition that the screwdriver can be used not just to turn screws but also as a pendulum weight.

Relatively little is known about how people solve insight problems. These problems are typically challenging to study in the laboratory with much depth, because it is difficult to ask people to describe the steps that they go through to solve them. On the other hand, there have been studies on the effects of taking a pause while working on a problem—called the "incubation effect." These pauses tend to increase the probability that the person will find the insight required to solve the problem.

No one has yet implemented a computational intelligence approach that changes the representation or that can recognize that a domino must cover a red and a black square. I expect that such a program is possible, but it would take a different approach to problem solving than has been attempted so far.

Path problems, like the hobbits and orcs problem or the Towers of Hanoi problem, all have the form of a search for a path through a series of possible states. Progress on path problems has been aided by understanding this part of their formal structure. In computational intelligence, progress has been aided by having faster computers that can compute more potential paths through the space and by heuristics that suggest which paths are more likely to be fruitful than others.

Insight problems, on the other hand, do not have the same kind of formal structure. They do not provide the representation or state space, and they may not have explicit rules for moving from one state to the next. In fact, they may have only two states or maybe three states (for example, wrong representation, right representation, solution). Humans can solve both path and insight problems, but they are so different from one another that understanding how path problems are solved is of little value to understanding how insight problems are solved. Computer scientists have extensively studied path problems but have done practically no work on insight problems.

One could argue that the real intelligence in setting up a computer system to solve a path problem like chess or go is the design of the state space, representation, methods for changing from one state to another, and maybe heuristics for selecting potential paths. Once these are created, there is not much to do except employ these tools. John McCarthy, an early pioneer in artificial intelligence, complained that once we understood how to solve an AI problem, it ceases to be considered intelligent.

Quirks of Human Intelligence

People do not seem ordinarily to pay a lot of attention to the formal parts of a problem, especially when making risky choices. For example, Tversky and Kahneman, as was mentioned, found that people would make different choices when presented with the same alternatives, depending on how these alternatives were described. We have already looked at the difference between early-bird discounts and late payment penalties.

Participants in one of their studies were asked to imagine that a new disease threatened the country, from which 600 people would be likely to die. They were further told that two programs had been proposed to treat these people. And they were asked to choose between two treatments. In the first version they were told:

> Treatment A will save 200 lives, whereas under Treatment B, there is a 33% chance of saving all 600 people and a 66% chance of saving no one.

Given this choice, 72% of the participants chose treatment A. Being certain to save 200 people was seen to be preferable to the chance that all 600 would be lost.

A second group was given a different version of the same choice:

> Under treatment A 400 people will die. Under treatment B, there is a 33% chance that no one will die, and a 66% chance that all 600 people will die.

In this second version, 22% of the participants chose Treatment A. Assuming that people believe that the numbers in each alternative are accurate, Treatment A is identical for both groups. Presumably, 600 people will die if no treatment is selected. In the first version, 200 of these people will be saved, meaning that 400 of them will die. In the second version, 400 people will die, meaning that 200 of them will be saved.

A rational decision maker should be indifferent to these two alternatives, yet the differences in people's preferences were substantial. The first version emphasized the positive aspects of the alternative, and the second one emphasized the negative. By a dramatic margin, people preferred the positive version.

It is worth noting that alternative B was also identical under both conditions. The expected number of people to survive under alternative B was also 200, but this alternative also included uncertainty. People preferred the certain outcome over the uncertain one when the certain one was framed in a positive tone and preferred the uncertain alternative when the certain one was framed in a negative tone. The frame or tone of the alternatives controlled the willingness of the participants to accept risk.

From a rational perspective, the effect of positive versus negative framing makes no sense. Formally, these alternatives are identical. One could say that this is an example of human foolishness rather than human intelligence. On the other hand, this error may tell us something important about how people make their decisions. Correct and incorrect decisions are both produced by the same brains/minds/cognitive processes.

Perhaps the apparent irrationality of the choices made in the treatment problem are due to limitations in the way that people can think about a problem in a short amount of time. People seem to have dramatic capabilities in some areas of cognition, but decidedly limited ones in others.

For example, people can recognize thousands of pictures and even details from those pictures. In one demonstration of this phenomenon, people were shown 10,000 images for a few seconds each. They were then tested by being shown two pictures, one of which they had seen and one that they had not seen. They could choose correctly in about 83% of these pairs (Standing, 1973).

On the other hand, Raymond Nickerson and Marilyn Adams (1979) asked people living in the United States to draw the front and back of a US penny. Try it, and see what you can come up with. Nickerson and Adams found that people were remarkably inept at remembering what was on a coin that they saw practically every day. Of the eight critical features that Nickerson and Adams identified, people included only about three. If you think it was because of the low value of the penny (it was worth more in the 1970s) or because we don't use coins much anymore, try recalling other common objects, such as a $1 or $20 bill or your credit card.

Unlike computers, people are relatively limited in what they can keep in active memory at one time. Digit spans were used in some early intelligence tests. In a test of digit spans, the examiner provides a set of random digits (for example, 5, 1, 3, 2, 4, 8, 9) to the person being tested, and the person is supposed to repeat them back immediately. Most healthy adults can repeat back about seven digits.

The typical limitation of about seven items is not limited to just numbers. In 1956, George Miller published a paper called "The Magical Number Seven, Plus or Minus Two." In it, he noted the wide range of memory and categories where people were limited to handling between five and nine items without making errors.

Miller was among the first cognitive psychologists to talk about memory chunks. People can adopt representations that allow them to expand the number of items that they can keep in mind. Chase and Ericsson found that one person could remember up to 81 digits after extended practice. This person, identified by his initials, SF, increased his memory span by organizing the digits into chunks that were related to familiar facts that he knew about, such as race times (he was an avid runner) or dates.

These and other psychological phenomena show that people have a complexity to their thinking and intellectual processes that is not always in their favor. People jump to conclusions. We are more easily persuaded by arguments that we prefer to be true or that are presented in one context or another. People do sometimes behave like computers, but more often, we are sloppy, inconsistent, and sometimes not too bright.

Daniel Kahneman describes the human mind as consisting of two systems, one that is fast, relatively inaccurate, and automatic. The other is slow, deliberate, and when it does finally reach a conclusion, more accurate. The first system, he said, is engaged when you see a picture and note that the person in it is angry and is likely to yell. The second system is engaged when you try to solve a multiplication problem like 17 × 32. The recognition of anger, in essence, pops into our mind without any obvious effort, but the math problem requires deliberate effort and maybe a pencil and paper (or a calculator).

Kahneman may be wrong in describing these as two separate systems. They may be part of a continuum of processes, but he is, I think, undoubtedly correct about the existence of these two kinds of processes in human cognition (and maybe ones in between). What he calls the second system is very close to what I call artificial intelligence. It involves deliberate, systematic efforts that require the use of cognitive inventions.

The bat-and-ball problem shows one way that the two kinds of process interact. Try to answer this one as quickly as you can. Let's say that to buy a bat and a ball costs $1.10. The bat costs $1.00 more than the ball. How much does the ball cost?

Most people's first response is to say that it costs 10 cents. On reflection, however, that cannot be right because then the total cost of the bat and ball would be $1.20, not $1.10. One dollar is only 90 cents more than 10 cents. The correct answer is that the ball costs 5 cents. Then the bat costs $1.05, which together add up to $1.10. The initial, automatic response can be overridden by a more deliberate analysis of the situation.

Computational intelligence has focused on the kind of work done by the deliberate system, but the automatic system may be just as or more important. And it may be more challenging to emulate in a computer. This rapid learning may sometimes result in inappropriate hasty generalizations (I always get cotton candy at the zoo), but it may also be an important tool in allowing people to learn many things without the huge number of examples that most machine learning systems require. A hasty generalization proceeds from one or a few examples to encompass, sometimes erroneously, a whole class of items. Ethnic prejudices, for example, often derive from a few examples (each of which may be the result of yet another reasoning fallacy, called "confirmation bias") and extend to large groups of people.

On the other hand, processes such as language learning rely on hasty generalization. A one-year-old child may know a few dozen words or maybe even 100, but a 12-year-old may know 50,000 to 75,000 words. That's a lot of learning going on over those 11 years. Many of those words will have been heard only once or twice. Learning them does not typically take deliberate effort (until the child starts to practice for tests like the SAT). Children just learn them as part of their daily experience from a small number of examples. They may misuse some of the rarer words, but they do have a concept of what they mean, even if that meaning does not match that of a dictionary writer. For example, a linguist friend of mine recounted a conversation she had with her son, who was worried about getting a potential treat later in the day if he behaved. He asked his mother if he was "being have." This was clearly a phrase that he had never heard before, but he had heard the instructions to "behave," and to "be good." He and his mother certainly talked about "being good." It was a natural extension to therefore think that it would appropriate to say "being have."

The deliberate system has its limitations as well. In one famous experiment, the researchers showed people a short video of two teams bouncing basketballs. They asked the participants to count the number of times one team bounced the ball and to ignore the other team. In the middle of this video, they had a gorilla, actually a woman dressed as a gorilla, walk through the scene, beat her chest, and walk off. The gorilla was visible for a full nine seconds in the video, but only about half the participants reported seeing her. Apparently, it took all of their cognitive capacity or attention to track the basketball bounces that they were counting, and there was none left to notice the gorilla.

Another example is called "change blindness." When people view a picture, they typically report that they see all of its parts. However, it is easy to show that their reports are incorrect. Change blindness is shown by displaying two pictures in alternating order, usually with a small time gap between them. The two pictures are different from one another, usually in rather unsubtle ways. For example, the two pictures may display an airliner. In one picture it has an engine under the left wing; in the other it does not. Even with deliberate effort, many people do not find the difference after many repetitions of the two pictures. [https://www.cse.iitk.ac.in/users/se367/10/presentation_local/Change%20Blindness.html]

People do not always see what they think they see. Their self-report of what they are doing and how they are doing it is not always a good indicator of what is actually going on when people perform cognitive tasks. If computers are going to emulate human intelligence, they will need to copy some, but perhaps not all, of the covert (Kahneman's System 1) processes that people engage. Yet these processes are generally inaccessible to deliberate description, so there is no obvious road map to implementing them. Nonetheless, they seem critical to human intelligence.

Quoc V. Le at Google (2012) deployed 16,000 CPU cores in a simulated neural network to process 10 million pictures over three days. After all of this training, their system was able to group pictures of human faces together. The pictures were carefully prepared for presentation to this system, all scaled to the same size (200 × 200 pixels) and selected to minimize duplicates. Compare this to the ability of people to recognize thousands of pictures after only a few seconds of presentation.

After all of this training, Le and his colleagues examined the outputs of the system and searched for one of the output simulated neurons that was best correlated with faces. They inferred that this neuron was the "face neuron" because it was the most active one on about 83% of the 13,026 faces in a test set of 37,000 pictures. In identifying this neuron, they used a method similar to that used in the 1950s, in analyses of frog visual neurons. Lettvin and his colleagues (1959) measured the output from frog visual neurons while they showed the frog different visual patterns until they found the pattern that produced the largest response in the neuron. They then labeled this neuron relative to that stimulus (for example, as a "bar" neuron).

In contrast to the Herculean effort Le's system required to cluster photographs, a child does not need anything like this level of effort to recognize human faces. Within an hour of being born, babies can recognize human faces relative to other scenes. Within hours of being born, infants learn to recognize their mother's face over a female stranger's (Bushnell, 2001; Bushnell, Sai, & Mullin, 1989; Pascalis, de Schonen, Morton, Deruelle, & Fabre-Grenet, 1995; Sai, 2005). They learn not just how to recognize a face; they recognize individual faces after just a few exposures.

By two months of age, babies can distinguish between novel pictures of general scenes and pictures that they have seen before. In other words, they show evidence of being able to classify pictures as novel versus familiar by the time that they are two months of age—and after less than 10 presentations of these pictures, not millions. At this point, we do not know very much about how babies learn so rapidly, but such rapid learning mechanisms may be an essential part of human intelligence.

If we are to design intelligent computers, we will need to know how humans learn what a face looks like so quickly. Infant brains are doing something different from what Le's computer network was doing, and identifying just what that difference is could be an essential part of creating general machine intelligence. Whatever babies do when they come to recognize faces, it is probably not "stochastic gradient descent with topographic independent component analysis."

Ultimately, it may be possible to accomplish similar goals with different means, but it seems to me that understanding this in the context of human intelligence is a critical part of accomplishing it in machines. The same brains that show the irrational quirks are also capable of such efficient learning, and it is likely that these characteristics are related.

Conclusion

A theory of artificial general intelligence would benefit strongly from a better understanding of just what intelligence is. Even though there is no clear consensus on what psychologists mean by intelligence, I think that it is clear that it involves more than formal problem-solving abilities.

An essential part of that theory of intelligence is the recognition that there are actually multiple kinds of problems, which appear to require multiple kinds of mechanisms to solve them. The problem of finding a representation that can be used to solve a problem is fundamentally different from navigating a path through a specified representation. Fast learning appears to be another essential feature of intelligence. Multiple iterations through thousands or millions of examples before something can be learned is a strong barrier to creating intelligence. Insight, particularly the ability to construct representations, is a critical skill for intelligence. We don't know a whole lot about how people generate these insights, but there are some clues, to be described in chapter 7. These are among the themes that we will return to through the rest of this book.

Resources

Anzai, Y., & Simon, H. A. (1979). The theory of learning by doing. *Psychological Review, 86,* 124–140.

Batchelder, W. H., & Alexander, G. E. (2012). Insight problem solving: A critical examination of the possibility of formal theory. *The Journal of Problem Solving, 5*(1), 56–100. doi:10.7771/1932-6246.1143; http://docs.lib.purdue.edu/cgi/viewcontent.cgi?article=1143&context=jps

Boole, G. (1854). *An investigation of the laws of thought on which are founded the mathematical theories of logic and probabilities.* New York, NY: Macmillan.

Brady, T. F., Konkle, T., Alvarez, G. A., & Oliva, A. (2008). Visual long-term memory has a massive storage capacity for object details. *Proceedings of the National Academy of Sciences of the United States of America, 105,* 14325–14329. doi:10.1073/pnas.0803390105; http://www.pnas.org/content/105/38/14325.full

Bushnell, I. W. R. (2001). Mother's face recognition in newborn infants: Learning and memory. *Infant and Child Development, 10,* 67–74. doi:10.1002/icd.248

Bushnell, I. W. R., Sai, F., & Mullin, J. T. (1989). Neonatal recognition of the mother's face. *British Journal of Developmental Psychology, 7,* 3–15. doi:10.1111/j.2044-835X.1989.tb00784.x

Chase, W. G., & Ericsson, K. A. (1982). Skill and working memory. In G. H. Bower (Ed.), *The psychology of learning and motivation* (Vol. 16, pp. 1–58). New York, NY: Academic Press.

Duncker, K. (1945). On problem solving. *Psychological Monographs, 58*(5), 1–113.

Ensmenger, N. (2011). Is chess the drosophila of artificial intelligence? A social history of an algorithm. *Social Studies of Science, 42*(1), 5–30. https://pdfs.semanticscholar.org/c9e7/3fc7ec81458057e6f96de1cba095e84a05c4.pdf

Ericsson, K. A., & Simon, H. A. (1980). Verbal reports as data. *Psychological Review, 87,* 215–251.

Gick, M. L., & McGarry, S. J. (1992). Learning from mistakes: Inducing analogous solution failures to a source problem produces later successes in analogical transfer. *Journal of Experimental Psychology: Learning, Memory, and Cognition, 18,* 623–639.

Jeffries, R., Polson, P. G., Razran, L., & Atwood, M. E. (1977). A process model for Missionaries-Cannibals and other river-crossing problems. *Cognitive Psychology, 9,* 412–440.

Kuhn, T. S. (1962). *The structure of scientific revolutions.* Chicago, IL: University of Chicago Press.

Lakatos, I., & Musgrave, A. (Eds.). (1970). *Criticism and the growth of knowledge.* Cambridge, UK: Cambridge University Press.

Le, Q. V., Ranzato, M., Monga, R., Devin, M., Chen, K., Corrado, . . . Ng, A. Y. (2012). Building high-level features using large scale unsupervised learning. In J. Langford & J. Pineau (Eds.), *Proceedings of the 29th International Conference on Machine Learning* (pp. 507–514). Madison, WI: Omnipress. https://static.googleusercontent.com/media/research.google.com/en//archive/unsupervised_icml2012.pdf

Legg, S., & Hutter, M. (2007). A collection of definitions of intelligence. arXiv. https://arxiv.org/pdf/0706.3639v1.pdf

Lettvin, J., Maturana, H., McCulloch, W., & Pitts, W. (1959). What the frog's eye tells the frog's brain. *Proceedings of the Institute of Radio Engineers, 47,* 1940–1959. https://hearingbrain.org/docs/letvin_ieee_1959.pdf

Loyd, S. (1914). *Sam Loyd's cyclopedia of 5000 puzzles, tricks & conundrums with answers.* New York, NY: Franklin Bigelow, The Morningside Press.

Maier, N. R. F. (1931). Reasoning and learning. *Psychological Review, 38,* 332–346.

Mill, J. S. (1967). On the definition of political economy; and on the method of philosophical investigation proper to it. In F. E. Mineka & D. N. Lindley (Eds.), *The later letters of John Stuart Mill, 1849–1873* (Vol. 4, pp. 309–339). Toronto, Ontario, Canada: University of Toronto Press. (Original work published 1836)

Miller, G. A. (1956). The magical number seven, plus or minus two: Some limits on our capacity for processing information. *Psychological Review, 63,* 81–97.

Nickerson, R. S., & Adams, M. J. (1979). Long-term memory for a common object. *Cognitive Psychology, 11,* 287–307.

Newell, A., and Simon, H. A. (1972). *Human problem solving.* Englewood Cliffs, NJ: Prentice-Hall.

Pascalis, O., de Schonen, S., Morton, J., Deruelle, C., & Fabre-Grenet, M. (1995). Mother's face recognition by neonates: A replication and an extension. *Infant Behavior and Development, 18,* 79–85. doi:10.1016/0163-6383(95)90009-8

Reynolds, G. D. (2015). Infant visual attention and object recognition. *Behavioural Brain Research, 285,* 34–43. doi:10.1016/j.bbr.2015.01.015; https://www.ncbi.nlm.nih.gov/pmc/articles/PMC4380660/

Sai, F. Z. (2005). The role of the mother's voice in developing mother's face preference: Evidence for intermodal perception at birth. *Infant and Child Development, 14,* 29–50. doi:10.1002/icd.376

Spearman, C. E. (1904). "General intelligence," objectively determined and measured. *American Journal of Psychology, 15,* 201–293. doi:10.2307/1412107; https://psychclassics.yorku.ca/Spearman/chap5.htm

Standing, L. (1973). Learning 10,000 pictures. *The Quarterly Journal of Experimental Psychology, 25,* 207–222.

Stanford Encyclopedia of Philosophy. (2016). Imre Lakatos. https://plato.stanford.edu/entries/lakatos/

Sternberg, R. J. (2003). An interview with Dr. Sternberg. In J. A. Plucker (Ed.), *Human intelligence: Historical influences, current controversies, teaching resources.* http://www.indiana.edu/~intell

Sternberg, R. J., & Detterman, D. K. (Eds.). (1986). *What is intelligence?* Norwood, NJ: Ablex.

Sundem, G. (2013). Study: Is complex problem solving distinct from IQ? https://www.psychologytoday.com/us/blog/brain-candy/201306/study-is-complex-problem-solving-distinct-iq

Tversky, A., & Kahneman, D. (1981). The framing of decisions and the psychology of choice. *Science, New Series, 211,* 453–458. http://links.jstor.org/sici?sici=0036-8075%2819810130%293%3A211%3A4481%3C453%3ATFODAT%3E2.0.CO%3B2-3; http://psych.hanover.edu/classes/cognition/papers/tversky81.pdf

Van Damme, E. (2010). Liquor filled chocolates. http://www.chefeddy.com/2010/09/liquor-filled-chocolates

Weisberg, R. W. (2015). Toward an integrated theory of insight in problem solving. *Thinking & Reasoning, 21,* 5–39. doi:10.1080/13546783.2014.886625

Wenke, D., Frensch, P. A., & Funke, J. (2005). Complex problem solving and intelligence: Empirical relation and causal direction. In R. J. Sternberg & J. E. Pretz (Eds.), *Cognition and intelligence: Identifying the mechanisms of the mind* (pp. 160–187). New York, NY: Cambridge University Press. http://cogprints.org/6626/1/Wenke_Frensch_Funke_CPS_2005.PDF

3 Physical Symbol Systems: The Symbolic Approach to Intelligence

In this chapter we discuss some of the early computational approaches to intelligence. The idea of the Turing machine, a general computational device, led to the notion of intelligence as a computational function. The Turing test provided a suggested means for evaluating whether a machine was able to execute a function similar to that of an intelligent human—in this case, to hold a conversation. From these two ideas grew the notion that intelligence is a symbol-manipulating process. This approach dominated the field of computational intelligence for about 30 years.

If defining intelligence precisely in the context of human intelligence is difficult, as we saw in the preceding chapter, it is even more difficult in the context of computational intelligence. In the early days following the Dartmouth workshop, artificial intelligence was taken to mean something like "the art of making machines do things that would require intelligence if done by men" (Minsky, 1968). In the original Dartmouth workshop proposal, the stated goal was to describe, with sufficient precision, every aspect of learning or intelligence so that a machine could be made to simulate it. Specifically, McCarthy and his colleagues thought to focus on investigating how to make machines use language, form abstractions and concepts, improve themselves, and solve the kinds of problems that only humans had previously been able to solve.

Herbert Simon, one of the attendees at the Dartmouth workshop, is quoted as saying soon after that meeting:

> It is not my aim to surprise or shock you—but the simplest way I can summarize is to say that there are now in the world machines that think, that learn and that create. Moreover, their ability to do these things is going to increase rapidly

> until—in a visible future—the range of problems they can handle will be coextensive with the range to which the human mind has been applied. (Simon & Newell, 1958)

Although there were attempts to build computational devices before this, the roots of this approach to machine intelligence are most directly attributable to seminal work by Alan Turing.

Turing Machines and the Turing Test

In 1937 Turing introduced a concept that came to be known as a Turing machine. A Turing machine is not a physical machine made of gears or transistors; rather it is an abstract computational idea. It is a mathematical description of a kind of ideal system that can implement any computable function. It is a model of computation that defines basic computational processes and can, in theory, be built to simulate the logic of any algorithm.

Conceptually, a Turing machine consists of a tape, marked out in squares or cells. Each cell may contain one symbol. Because the Turing machine is conceptual, the tape can be assumed to be of infinite length. The machine has a "head," which can read or write a symbol in one of the cells, and a "state register," which holds information about the current state of the machine. The symbols that the machine reads from the tape can change the machine's state among a finite number of potential states. The state is affected by the symbols that have been read so far. So the state acts as a kind of memory. Depending on the symbol in the currently read cell, the machine's state, and a finite table of rules, the machine can change state, write a symbol into the current cell, or move to a different cell on the tape (Turing, 1965/1936).

For example, given that the machine is currently in state 57 and the cell under the read/write head contains a 0, the machine may move one cell to the right, change into state 128, and write the symbol 1 in this new cell. The states, the symbols, and the rules are all finite, but given that the tape has no practical limit and that time is not limited, a Turing machine can conceptually have infinite capacity.

With this abstract machine, Turing was able to answer fundamental questions about computability. Modern computers are generalizations and physical instantiations of Turing's original proposal. They have more complex rule sets, they have multiple registers (not just one) where they can keep temporary information, and they have random access memory (as opposed to the sequential memory of an even infinite tape).

One of the reasons that the Turing machine is so important to computational intelligence is that if intelligence is computable, then it would be computable by a Turing machine. If it can be computed by a Turing machine, then it would be computable by any machine that was equivalent to a Turing machine. Turing believed that intelligence was, in fact, a computable function, and he proposed a test that would evaluate this belief. This test came to be called the "Turing test."

The word "computable," especially in the context of a "computable function," has a special meaning in computer science. Computable functions are algorithms. They consist of a step-by-step procedure that takes an input and produces a definite output. According to the Church–Turing thesis (named after Alonzo Church, whose paper on the topic came out in the same year as Turing's paper on computability), a computable function is one that could be implemented on a calculation device that had access to an unlimited amount of time and an unlimited amount of storage space (Turing's infinite tape). Operations that require a lot of memory or that take a long time may still be computable as long as the process could be completed if the system had sufficient resources. This notion of computable is concerned with the theoretical limits of computation, not with the practical limits.

To be computable, the procedure must be specifiable with exact instructions, such as in a computer program. Given a set of inputs, it must produce an output after a definite number of steps, and the output must be verifiable.

Not all functions are necessarily computable. Even if a computer can execute some program, that does not mean that the program is computable in the Church–Turing sense. For example, Turing proved that the so-called halting problem was an example of a function that is not computable, even with unlimited resources.

The halting problem is a decision problem: determine from a description of a computer program and an input whether the program will finish and produce an output or will run forever. For really simple programs, it is fairly easy to make this decision, but for programs with some complexity, Turing showed that it was impossible to reliably make this decision.

If we run the program and it stops within a short time, that's great. We can decide that the program does, indeed, finish. But if we run the program and it does not stop, is that because it will never complete or because we simply have not run it long enough? If we run it longer, does that provide any better evidence that it will not complete? For all we know, the very next computational step will lead to its completion, or maybe not. Some algorithms take a very long time to complete, and some may never complete. We can prove that some programs will halt and produce an output, but we cannot, in general, prove that it will never finish. It is impossible to prove a negative.

If we cannot prove that it will complete, then we say that the function is not decidable. Notice the relationship between decidability and proof. The idea is that we must not only make a decision but verify that this decision is correct. Turing proved that a general algorithm addressing the halting problem for all combinations of programs and inputs cannot be written.

A function, in the computer science sense, takes an input and produces a specific output. Even if intelligence is a function, it may not be a computable function. It may not be a provably decidable function. It could still be executed by a computer, but we may never be able to prove that its answer is definitely correct. In fact, I argue that intelligence is not a decidable function in this sense, but it can still be implemented by brains and computers. Intelligence, in other words, is not an algorithm, but is, perhaps, a different form of computation.

Turing argued that the work of one of his machines can be used to do the work of any other computer. We would simply have to include a description of the machine we want to emulate as part of its tape. Presumably, if we had a proper description of the brain, we could include that description on the tape and use a Turing machine to emulate it.

From this claim, it is easy to see why Turing machines were so important to computation and to computational intelligence specifically. They introduced the notion of equivalent machines. Any two computers that compute the same function are, in this conceptualization, equivalent machines, and it does not matter whether they were made of cogs and wheels, mercury delay lines (as Turing proposed), vacuum tubes, integrated circuits, or perhaps even brains. Same function, equivalent machines.

The idea of a Turing machine also suggested that a machine, by manipulating symbols, could implement any conceivable act of formal reasoning.

If it turned out that human thought could be implemented as acts of formal reasoning, as Newell and Simon claimed, then the Turing machine would be a proof that a computer could, in fact, duplicate it and be an equivalent machine to human thought. Much of the history of computational intelligence research, at least through the 1980s, could be described as an attempt to show that formal reasoning was enough to implement the equivalent of human intelligence. Theorem proving, the topic that Simon and Newell focused on, is, of course, a quintessential example of formal reasoning.

Turing's work was an essential part of practically every computational advance over the next 50 years or more. His ideas of equivalent machines as those that compute the same function, and his concept of the Turing test (described next), were also central to our understanding of computational intelligence. It was not until many years later that it began to be apparent that computability was also a limitation that computational intelligence needed to overcome.

Turing proposed his "imitation game" test in 1950 in the context of the question of whether a computer could be said to think. The concept of thinking is somewhat amorphous, so Turing asked, instead, whether a computer might do well in an imitation game. Essentially, he drew on his notion of equivalent machines to propose a test of machine intelligence. If a machine could hold a conversation that was indistinguishable from a conversation held with a human, then we would be compelled to consider that computer intelligent. If it executes the conversation function in a way that is indistinguishable from the way a human executes that function, then Turing would call it an equivalent machine.

The test assumes that conversation is the right measure of intelligence. But beyond that, if two systems are indistinguishable in that function, then we should not attribute properties to one that we do not attribute to the other.

Intelligence does not depend on the ability of the computer or the person to render speech, so the assessor, in Turing's proposal, would communicate with the candidate (the computer or the person) by typing and reading written responses. It is the form of the conversation, not its physical channel, that is important. If the assessor cannot tell whether the written responses, perhaps presented on a computer terminal (in those days, on a Teletype), were coming from a person or from a computer, then the computer has passed.

Turing anticipated modern machine learning in a presentation he gave to the London Mathematical Society in 1947 (Turing, 1947/1986). He supposed that one could set up a Turing machine or its equivalent digital computer with a table of instructions and the ability to modify those instructions. He supposed that after some amount of operation, the computer would have modified its instructions beyond recognition. He compared this computer to a student who had learned initially from his teacher but then added much more of his own work. "When this happens I feel that one is obliged to regard the machine as showing intelligence."

Once a computer can modify its own instructions, then any question about the computability of the function that computer is running is off the table. The modifications of its computational pattern mean that we cannot predict with assurance whether it will produce a specific output, given a specific input. The concept of computability includes the idea of a definite process with a finite number of steps. Once a computer program can modify its own operation, it is no longer guaranteed to be following a finite set of steps. The computer may contain any number of computable functions that allow it to operate, but the overall function of intelligence is not, by this definition, computable. The process fails to be computable not because we cannot decide its status but because it is no longer executing a specific effective procedure.

If we give up the certainty of the Church–Turing effective procedure, we do not have to give up the idea that a computer is computing something effective in the more colloquial sense. Turing recognized this in his 1947 report to the London Mathematical Society ". . . if a machine is expected to be infallible, it cannot also be intelligent. There are several theorems which say almost exactly that."

A few years after Turing proposed his models of computability, Warren McCulloch and Walter Pitts (1943) showed how the brain could also be a Turing-equivalent machine. They showed that their conceptualization of neurons could be organized to compute the basic logical functions. "Because of the all-or-none character of nervous activity, neural events and the relation among them can be treated by means of propositional logic" (McCulloch & Pitts, 1943, p. 115). As you might expect, from the vantage of more than 70 years of research into neurons and their activity, their idea of how neurons represent mental activity was oversimplified; it nevertheless represented a radical departure for thinking about the mind and brain—one that was highly influential in the development of artificial intelligence.

McCulloch and Pitts's hypothesis was among the first to propose a formal theory of mind (that the mind was a Turing-equivalent machine) and among the first to talk about neural networks and computation. The basic idea is that all logical relations can be represented as some combination of the logical operations AND, OR, and NOT. McCulloch and Pitts showed how neurons could, in fact, perform these logical operations. Therefore, some organization of neurons could be the equivalent of a Turing machine.

The Dartmouth Summer Workshop (1956)

The idea of brains as Turing-equivalent machines was also one of the main motivations for John McCarthy to organize the 1956 Dartmouth summer workshop. Like Turing, McCarthy was interested in the mathematical properties of intelligence. He had done some work on the application of mathematics to commonsense reasoning and had worked for Claude Shannon, who is often called the father of information theory. McCarthy's background in mathematics, particularly mathematical logic, and his interest in the brain inspired him to organize a workshop, which already been mentioned several times, directed at "the conjecture that every aspect of learning or any other feature of intelligence can in principle be so precisely described that a machine can be made to simulate it. An attempt will be made to find how to make machines use language, form abstractions and concepts, solve kinds of problems now reserved for humans, and improve themselves."

Among the topics that they proposed to consider were:

- "How can a computer be programmed to use a language?" They noted that "a large part of human thought consists of manipulating words according to rules of reasoning and rules of conjecture." Can we write a program that allows statements to imply others?
- Neural networks. "How can a set of (hypothetical) neurons be arranged so as to form concepts[?]"
- Machine learning. "Probably a truly intelligent machine will carry out activities which may best be described as self-improvement."

- Creativity and randomness. "The difference between creative thinking and unimaginative competent thinking lies in the injection of a some [*sic*] randomness" but this randomness must be guided by intuition.

The participants in this workshop, John McCarthy, Marvin L. Minsky, Nathaniel Rochester, and Claude Shannon, plus Allen Newell, Arthur Samuel, Oliver Selfridge, and Herbert Simon, went on to write computer programs that could do things like play checkers, solve algebra word problems, prove logic theorems, and converse in English. The attendees of this conference were a veritable who's who of the people who would become important in the future of computational intelligence.

Herbert Simon and Allen Newell discussed their program (written with John Shaw), the Logic Theorist, at the Dartmouth workshop. In fact, they were the only participants who had a working program demonstrating some aspects of artificial intelligence.

Their program was designed to prove mathematical theorems like those in Bertrand Russell and Alfred Whitehead's book, *Principia Mathematica*. And it was able to prove *a substantial number of them*. Simon saw this work as a major accomplishment (see chapter 1) in part because it used the computer symbolically, not just as a mathematical calculator.

The Logic Theorist was an attempt to take advantage of the power of symbol manipulation techniques. Newell and Simon were helped to achieve their progress because Russell and Whitehead had already expressed the theorems in their book in a precise symbolic form. The key idea of the Logic Theorist was that any problem that could expressed as well-formed formulas of a certain type could be solved by their approach.

Like many of the computational intelligence approaches that came after them, Newell and Simon argued that intelligence could be represented as a formal system and a computer program could be written that would be able to navigate that formal system. Like checkers-, chess-, and go-playing computers would eventually do, their approach was the equivalent of a graph that connected axioms to a conclusion, usually through several intermediate steps. The role of the computer was to search through this graph for the set of steps that would lead from the axioms to the conclusion, thereby proving the theorem. Despite the limited computational power available in the middle of the twentieth century, their approach was able to solve a number of problems, which led them to expect that a more complete solution would be available with only a little additional work and a little more computational power.

Newell and Simon's presentation received a lukewarm reception at the Dartmouth conference, but its introduction of the concept of reasoning as search had a profound and lasting effect on computational intelligence. Their method started with an initial hypothesis. Each branch was then a deduction from this hypothesis based on the rules of inference described in *Principia*. The set of deductions that led to the goal was the proof of the theorem. You can describe this path as a selection among the branches of a logic tree.

The Logic Theorist also used heuristics to select which paths to try. The notion of a heuristic was introduced by George Polya, with whom Newell had studied at Stanford, in the context of proving theorems, so it was a small step to use it in this context as well. The use of heuristics is critical to most kinds of problem solving, not just artificial intelligence. Heuristics are essential, particularly when there are too many possible paths or branches to follow. The heuristics limit the number of branches that need to be followed by selecting the ones most likely to be useful.

Newell, Simon, and Shaw's later program, the General Problem Solver, excelled at logical and geometric proofs. It separated its knowledge of problems, the rules, from the means of solving them, and so became a generic problem-solving engine. It could solve the Towers of Hanoi problem and play chess when these problems were expressed as well-formed formulas.

One heuristic it used for dealing with the large number of possible paths from the axioms to its conclusion was to choose the branch that would bring it closer to its desired conclusion. This heuristic can be called "hill climbing" because it chose the option that would take it further toward the top of the hill, the intended conclusion.

Representation

Newell and Simon later elaborated their notion of computational problem solving into the physical symbol system hypothesis (1976): "A physical symbol system has the necessary and sufficient means for general intelligent action." "Necessary" means that without a physical symbol system, one cannot achieve general intelligent action. "Sufficient" means that having a physical symbol system is all one needs to achieve general intelligent action.

A physical symbol system is one that takes physical symbols (such as marks on paper or symbols encoded into a computer) and combines them into structures or expressions. It then manipulates those expressions to produce new expressions. The symbols are physical objects that denote, represent, or stand for other objects. They may be virtual, they may be states of some physical system (such as an electrical charge on an integrated circuit, or a magnetic domain on a disk), but they are still physical. They are physical things that stand for other things. The physical symbol system, including its symbols and the rules for manipulating them, mapping them to other symbols, combining them into expressions, and transforming those expressions, is argued to be the basis for intelligence.

The physical symbol system hypothesis implies that computers can be intelligent if we just give them appropriate symbol-processing programs. Human thinking is symbol manipulation, under this hypothesis, and anything capable of such symbol manipulation is capable of being intelligent.

The physical symbol system hypothesis implies that there is a set of basic, primitive, irreducible symbols that form the core of the system, for example, axioms. Other symbols are defined by expressions using these basic symbols. The physical symbol system hypothesis casts intelligence as a formal system. This approach follows directly from Newell and Simon's original endeavor to try to prove all of mathematics from a set of axioms following Russell and Whitehead.

There are many criticisms of the physical symbol system hypothesis (see Nilsson, 2007). Among these is the question of whether it means anything more than that knowledge and intelligence can be digitized. If the ones and zeros of a computer comprise symbols in Newell and Simon's sense, then there is not much to their claim. Ones and zeros in a computer, for example, don't by themselves represent anything in the world. However, expressions including digits can represent things in the world. The number 438 (in binary: 110110110), for example, could stand for the number of calories in a piece of apple pie or for item #438 on a trucking manifest. Whether encoded as ones and zeros or as decimal digits, the mere presence of a number does not provide a symbolic representation without a process to specify its representational relation. Symbols have to stand for something to be symbols, they cannot just be isolated tokens. External reference is necessary, and external reference is not simply a formal process in the way that deductions from observations can be a formal process. The need for external reference is exactly the same problem that we considered in the context of logical positivism in the previous chapter. The problem of how symbols acquire their referential status is called the "symbol grounding process."

"Symbol grounding" is the process by which a symbol gets related to something outside of itself. Formal systems cannot infallibly connect symbols with their meanings. However, the operations that are appropriately performed on a symbol or expression depend on the "meaning" of that symbol. A symbol representing an item on a manifest might be added to or deleted from a list, but a symbol that represented the number of calories in a piece of pie might be added to a daily calorie budget. Not all operations make sense for all symbols.

The formal system is not entirely separable from the semantics (meaning) of the symbol. Meaning contaminates the symbol system, so a symbol system cannot be sufficient to produce intelligence. Newell and Simon could gloss over the symbol-grounding problem when proving the theorems of *Principia* because theorem proving in that context is already a purely formal system. Solving other kinds of problems, however, does involve issues of symbol grounding.

Language is a quintessential symbol system that Newell and Simon took as their inspiration for the physical symbol system hypothesis. But language does not work as a purely formal system. Language too suffers from the intrusion of meaning to affect what can be done with the symbols. One example of this kind of intrusion is called the "locative alternation." Some verbs can be used in multiple sentence structures without changing their meaning. For example, the following two sentences mean essentially the same thing:

- Leo hit the fence with a stick.
- Leo hit a stick against the fence.

But the next two sentences do not mean the same thing, even though the forms of the second pair are identical with forms of the first. The only difference is the use of the word "hit" versus the word "broke":

- Leo broke the fence with a stick.
- Leo broke a stick against the fence.

In the second pair, it is the fence that broke in the first sentence but the stick that broke in the second.

Hitting is a so-called locative verb—it puts the stick at the fence—but breaking is not a locative verb. Putting or removing verbs can take this alternate verb phrase form, but nonlocative verbs cannot. The meaning of the verb affects the sentence structures in which it can be used to convey a certain meaning. The world and the representational system cannot be rigidly separated. We will return to the symbol grounding problem shortly.

Criticisms like these further indicate that formal symbol systems are not adequate for computational intelligence. It is possible to argue that computers only process ungrounded arbitrary symbols, but there is no reason to think that the brain has any better way to ground symbols, like those in expressions such as "John's wife," "the apples from Washington," or "unicorns do not exist." Those symbols are no more formally grounded than the symbols in a computer. Rather, the meaning of the symbols comes from how those symbols are used. Even without a guarantee that the usage is correct, constraints on how symbols are used provide their meaning.

The approach to rule based intelligence inherent in the physical symbol system hypothesis, despite its problems with symbol grounding, continued to have a profound effect on computational intelligence in the form of so-called expert systems.

In the 1970s, computer scientists, particularly Edward Feigenbaum and his colleagues, began to investigate the possibility of building practical reasoning engines, which came to be called expert systems. Expert systems are intended to solve complex problems by emulating the reasoning of human experts, where this reasoning is represented by if-then rules. For example, a medical diagnosis system might have a rule of the form "If the patient has a rash, then consider whether the patient also has other symptoms of measles." They were not intended as models of general artificial intelligence. Rather, they were special-purpose applications that address specific problems.

Expert systems are strongly in the tradition of physical symbol systems. Feigenbaum was a student of Simon's. The rules of an expert system were ordinarily written down explicitly by "knowledge engineers" working in collaboration with subject matter experts. One of the first of these systems was DENDRAL, a program intended to help chemists identify unknown organic molecules from their mass spectrographs.

A mass spectrograph is a device that breaks complex chemical compounds into their elements and describes the relative amount of each element. It is instrumental in uncovering the chemical makeup of the compound and an important step in identifying its chemical structure. Even if we know the chemical elements that make up a compound, there are still many possible ways that those elements can be organized. DENDRAL was intended to use expert knowledge about the organization of chemicals to heuristically limit the range of possible organizations that had to be examined.

As a program, DENDRAL enforced a strict separation between its knowledge base (information about chemistry) and its inference engine. New content could be supplied to the system without having to change any code. Knowledge could change, but inference methods would remain the same when moving from one kind of task to another.

Feigenbaum and his colleagues attributed the success of DENDRAL and other expert systems to the knowledge base it contained (its set of facts), rather than to the system's reasoning power: "A system exhibits intelligent understanding and action at a high level of competence primarily because of the specific knowledge that it can bring to bear: the concepts, facts, representations, methods, models, metaphors, and heuristics about its domain of endeavor."

A later program called MYCIN was intended to diagnose blood diseases. Using about 450 rules, acquired by knowledge engineers through extensive interviews with experts, MYCIN was able to perform as well as some medical experts and better than most junior physicians.

DENDRAL, MYCIN, and similar programs were the first effective AI programs that accomplished more than solving toy problems. By 1982, expert systems were in commercial operation at Digital Equipment Corporation and many other companies. They translated the research problems of artificial intelligence into practical systems that solved real-world problems.

In the late 1980s I wrote an expert system to diagnose closed head injuries for what was then the US Veterans Administration. Closed head injuries occur, for example, when a person is struck in the head hard enough to cause an injury, such as a concussion, but not so hard as to crack the skull. Such injuries are common in battles (and, it turns out, in the NFL), and it is often difficult to get the victims to a place where they can be diagnosed by an expert. The rules for this system were based on extensive interviews with a neuropsychologist expert. The interviews allowed me to capture the rules that the expert uses. Once written down, they could be easily encoded into the system's knowledge base.

In an expert system, the rules are explicitly coded into the system's knowledge base by the knowledge engineer. A huge advance in artificial intelligence occurred when methods were later developed to allow the computer to learn its own rules. We will return to a consideration of the implementation of expert systems in the next chapter.

The emergence of expert systems had three valuable consequences for computational intelligence. First, they showed that at least some problems could be addressed by an embodiment of a physical symbol system. Second, they took AI out of the laboratory into solving real-world problems. Third, they did not pretend to solve the problem of general intelligence but limited their scope to specific problems using general methods. Arguably, the shift from general intelligence to the solution of specific problems is the most important and long-lasting of these consequences.

This shift from general intelligence, as described in the proposal for the Dartmouth workshop, to narrow intelligence is a striking pivot. The unrealistic promises of artificial general intelligence, and their ultimate disappointment, prevented the field from achieving long-term stability. Each breakthrough seemed promising, but progress was ultimately limited along with the funding when general intelligence failed to materialize. Once expert systems began to be deployed in real business settings, however, it became easy to see clear progress with demonstrable value.

Within a few short years most researchers in computational intelligence had abandoned any overt attempts to build a general enough machine to be capable of any task that could be performed by a human. They focused instead on building "dumb specialists in small domains" (Minsky, 1996). Success in narrow computational intelligence has been dramatic and remarkable, especially over the last decade or so.

But the goal of an artificial general intelligence has not been completely abandoned. It is still in many senses the holy grail of computer science. The bulk of the work, however, has concerned solving specific problems where progress is more immediate and more easily assessed.

Early AI systems, including early expert systems, suffered from the lack of computing power. Ross Quillian (1969) had a demonstration of a language understanding system, but it had a vocabulary of only 20 words because that was all that would fit in memory. His program, the Teachable Language

Comprehender, contained a semantic network, another fundamental innovation, that represented facts about its world and the connections between these facts. The network contained units that represented words or concepts and properties that represented characteristics of those words or concepts. Quillian's 1969 paper on the Teachable Language Comprehender was little more than a promissory note or a sketch of a solution. To be really useful, it would require a much larger memory of basic facts and relations, better understanding of sentence structure, and stronger inference rules, but it was a kind of proof of concept.

Lack of computer resources continues to hamper AI work, but Moore's law (roughly that computer capacity would double about every 18 months) and other progress in computing have brought us a long way from a time when 640,000 bytes (640k or 640,000 characters) was the norm for the amount of memory a system could hold. There is more computer power currently sitting on my wrist than there was in the college computer system I first learned to program in the 1970s. In 1976, Hans Moravec observed that computers would need millions of times more power than was currently available to actually show artificial intelligence.

The number of possible paths that would need to be considered for useful computational intelligence was also a formidable barrier to success. Beyond simple toy problems, the combination of all possible solutions meant that it would take an unreasonable amount of space to contain the alternatives and an unreasonable amount of time to try them all. Some heuristics were available to reduce the space and time burden, but they could not eliminate the burden. Since then, progress in computational intelligence has come from improvements in computational capacity and from the invention/discovery of new heuristics. We return to this idea near the end of this chapter.

Attacks on the physical symbol system hypothesis became widespread in the late 1970s and early 1980s. Much of the criticism came from philosophers, but some came from computer scientists as well. There were two general classes of criticism. First, the symbol grounding problem, as mentioned earlier, is the argument that symbols need some kind of connection in the real world in order to be useful as the basis of intelligence. The second is that the kind of computation described by the physical symbol system hypothesis is not the right kind of computation to create intelligence.

The symbol grounding problem is that the symbols being manipulated by the physical symbol system are just squiggles and strokes if they are written down, or ones and zeros if they are encoded in the computer. The system has no idea, according to the argument, that the symbols stand for something, that they are symbols of something. The word "apples" is just some marks on paper for a physical symbol system, but for an intelligent person, that symbol "refers" to a fruit. Real understanding, the argument goes, has to involve more than just manipulating symbols; it has to refer to the real world. Real understanding requires more than just a formal symbol-manipulating system.

The nature of how symbols refer for anyone, let alone a computer, is a difficult philosophical problem that we are not going to solve here. Words can be ambiguous ("Apple" refers to a company as well). Meaning is not the same thing as reference (a traditional example is the meaning of the phrase "Golden Mountain" in the sentence, "the Golden Mountain does not exist" cannot refer to anything because it does not exist). But setting those problems aside, it can still be said that a physical symbol system contains symbols and rules, without a connection to anything outside of itself. Intelligence, the argument goes, requires that outside connection.

John Searle proposed a thought experiment that he called the Chinese room. Imagine that you are in a room with bushel baskets of Chinese characters. You do not speak or read a word of Chinese, but you have a complete rule book and an unlimited set of physical Chinese symbols. A human Chinese speaker outside the room pushes Chinese characters, symbols, into the room by one window. The person outside the room actually does understand Chinese, but the person inside the room does not.

When you receive the symbols, you consult your rule book and push other symbols out of a second window. Because the rule book is sufficiently complete, every response you make is the conversationally appropriate response to the input that you received. This system, in other words, passes the Turing test.

As far as the Chinese speaker is concerned, she is having a conversation, although with a funny way of passing the symbols, in Chinese. But you, inside the room, still understand nothing of what is going on except to consult the rule book and do what it says. The Chinese room is a perfect embodiment of a physical symbol system, yet, according to Searle, it understands nothing so it cannot be intelligent.

It may be fair to say that the person inside the room still does not understand Chinese, but there is no person inside your head. We say that you understand Chinese when you can hold a conversation in it. None of your neurons, the cells that do the processing in your brain, understand Chinese, but you do. The room may not understand Chinese, but the process operating in that room surely behaves as if it does. The rule book, the symbols, and the implementation of the rules in the rule book together behave as if they understand Chinese, and maybe they do.

If I were to look inside the room, I might see you shuffling Chinese symbols around, but I cannot look inside your head in the same way. If I could, I might see neurons firing in various patterns, but that would not tell me directly whether you understand Chinese. I cannot observe Chinese understanding in your head any more than I can observe Chinese understanding by looking into the Chinese room. We attribute understanding to Chinese-speaking people with no more evidence that they understand than we get from the system in the Chinese room. I fail to see the difference, but for Searle, that difference is fundamental.

The whole thought experiment rests on the idea that the complete properties of language can be written in a rule book. I find that hypothesis implausible as well. Fundamentally, Searle's argument is that computers are capable of only rule following and rule following is not enough to create intelligence. Brains do more than follow rules; they understand. It is not clear, however, just what *understanding* is, such that a brain could have it, but a computer could not. In the next chapter, we return to consideration of symbol grounding in the context of expert systems.

The other main criticism of the physical symbol system hypothesis focuses on the kind of computation that computers are capable of. One version of this argument is that intelligence rests on nonsymbolic or analog processes.

A formal system, like that proposed for the physical symbol system hypothesis, takes expressions "written" in one set of symbols and produces other expressions involving these or different symbols. It assumes, though, that there is some set of basic symbols in which these expressions can be written. These basic symbols are considered to be "atomic" units. An atomic unit is the basic building block, and it cannot be reduced any further. In language, the most atomic unit of meaning is the morpheme. A word such as "unfriendly" can be broken down into three morphemes: "un," meaning roughly "not"; "friend," meaning roughly "companion"; and "ly," indicating that this word is an adverb.

Atomic symbols can be combined into expressions, analogous to phrases or sentences. The symbol for the fruit plum means "plum." It does not mean "apricot." But now we have a new fruit, a pluot, which is a combination of a plum and an apricot. Do we now need a new atomic symbol for pluots? Just how many atomic symbols do we need? How do we decide which symbols are atomic and which are defined by expressions? Can we identify the basic atomic units or axioms of intelligence? The prospects for identifying such basic elemental symbols are dim. The effort to axiomatize mathematics failed (Gödel, 1931/1992), and there is no reason to think that it should succeed in intelligence.

"Delicious apples" is an expression for a kind of apple. In theory, it consists of the atomic symbol for "delicious" and the atomic symbol for "apple." Neither "horse apples" nor "wax apples" are actually a kind of apple at all. So even if "apple" is a symbol, it is not actually an atomic symbol in that it does not always have the same, unchanging meaning. Its meaning changes as a function of the context in which it appears.

Attempts to derive the set of axioms or basic atomic facts about the world have not resulted in any kind of success. Facts seem to depend on context, and there seem always to be categories that are between any two categories we might come up with.

Douglas Lenat has been working on the CYC system for more than 30 years. CYC is intended to assemble a comprehensive representation of everyday commonsense knowledge. Soon after the project started, in 1986, Lenat estimated that the project would comprise 250,000 rules and take 350 person years of effort. An open version of CYC was available for a while. In its latest iteration, it describes 239,000 concepts with over 2 million facts about those concepts. The commercial version of the system contains additional base facts and is still not very successful at commonsense reasoning. Knowledge of the real world is very difficult to reduce to a set of axioms and inferences drawn from them.

Intelligence in the natural world seems to require some nonsymbolic processes. By nonsymbolic, we usually mean some kind of analog signal and the processing of that signal. It is not the presence or absence of a sound, for example, that serves as a symbol, but the amplitude (loudness) of the sound and the pattern of frequency changes over time that carry the information needed to hold a conversation. Both amplitude and frequency can vary continuously, making them analog properties.

To be sure, we can represent any analog property using symbols, like the digits zero through nine for increasing loudness (see the above discussion of the digital version of the physical symbol system hypothesis), but these symbols emulate or approximate the analog process. Loudness does not occur in, say, 10 discrete categories. Film and vinyl records are examples of analog recordings. CDs, DVDs, and Blu-rays are examples of digital approximations to those analog signals.

The intellectual tasks, such as chess playing, chemical structure analysis, and calculus are relatively easy to perform with a computer. Much harder are the kinds of activities that even a one-year-old human or a rat could do. This is called Moravec's paradox, which I think should better be called Moravec's irony. The things that people find difficult are relatively easy to do with a computer (checkers playing, reasoning, logic), but the things that people find easy, automatic, or even unconscious have been a challenge for computers. Building a robot that can fold a family's laundry, for example, turns out to be very difficult. Two companies have promised laundry-folding machines, but neither is currently on the market and both need help at some point in the process (Lee, 2018). We're not there yet, though one of these companies has apparently made some progress (Lee, 2019).

Recall the prime conjecture of the Dartmouth AI conference "that every aspect of learning or any other feature of intelligence can in principle be so precisely described that a machine can be made to simulate it." Where computational intelligence has had the most success is precisely in those processes that are relatively easy to describe in precise detail. Where it has had less success is in those tasks that are difficult to describe in detail. Accessibility to description is the key to computational success, at least until recently.

Human brains have evolved mechanisms over millions of years that let us perform basic sensorimotor functions. We catch balls, we recognize faces, we judge distance, all seemingly without effort. On the other hand, intellectual activities are a very recent development. We can perform these tasks with much effort and often a lot of training, but we should be suspicious if we think that these capacities are what makes intelligence, rather than that intelligence makes those capacities possible. Reasoning and similar processes may be merely the tip of the iceberg of intelligence, resting on a base that is much less formal and much less structured, but essential nonetheless.

Neurons are not fundamentally symbolic. Although the human mind can simulate a physical symbol system—for example—with effort, we can try to think logically. However, the evidence seems to indicate that the mind is inherently fuzzy. We have to go to school to learn logic, but logic does not always save us from making the wrong decisions. Consider the framing effect discussed in the preceding chapter—Kahneman and Tversky's finding that people make different decisions depending on how the question is presented or framed. If it is the case that humans only simulate physical symbol systems, but our underlying processes are fuzzy and irrational, then that would falsify the physical symbol system hypothesis. Boole and Simon were wrong; physical symbol systems are neither the basis for nor are they necessary for intelligent thought.

Scale was another serious challenge to the approach suggested by the physical symbol system hypothesis. Playing the game go was a serious challenge for computers because of all of the possible combinations of moves that must be considered. As the number of basic facts increases, the number of ways those facts can be combined explodes. This situation is called "combinatoric explosion."

Consider a simplified chess-playing program, and assume that at each move there is an average of just 20 possible choices. Assume further that we want to choose a move that will lead us toward success, so we try to look ahead to predict what the situation will be 15 moves ahead. Following each of the 20 possible moves could lead to 20 possible next moves, so looking even two moves ahead involves evaluating 400 possible alternatives. Looking 15 moves ahead will require us to examine 20 to the power of 15 moves, which is about 3.3×10^{19} (33,000,000,000,000,000,000, or 33 quintillion) possible moves. If our program could evaluate a billion moves a second, it would still take 90 million hours (10,000 years) to make a single move.

Most of the computational intelligence demonstration projects in the 1960s through the 1980s focused on simple toy problems because they did not have the computer resources to deal with more realistic ones. Recall that Quillian's language comprehender only managed a vocabulary of about 20 words. The investigators working on these problems thought that faster processors and more memory would enable them to easily increase the scope of their representations, but, as it turns out, they wildly underestimated the need, because they failed to properly account for the combinatorics of expanding their programs. Theorem-proving programs, for example, failed at proving theorems with more than a few tens of facts.

Obviously, our chess-playing program will need some method to limit the number of combinations it considers, and that is where heuristics come into play. When an exhaustive analysis is infeasible, we have to find some manageable analysis that will get the problem solution approximately correct most of the time. Some developer has to have enough specialized knowledge of the problem and enough insight to generate sensible heuristics.

The difficulty of looking ahead in a chess problem shows that not all path problems are straightforward to solve, even if they are straightforward to describe. There is another class of problems that are even more intractable, even though they sound like something that should be easy to solve. One of these is the knapsack problem. Given a knapsack with a certain capacity, for example, a maximum weight that it can carry, and a set of objects, each of which has a known weight and a known value, determine the items to include so that the total weight is less than the carrying capacity of the knapsack and the total value of the contents is as high as possible.

With only a small knapsack, the problem is not terribly difficult, but the framework of the knapsack problem can be applied in a large number of situations. For example, professional sports teams may have a limit on the total amount of money the team can spend on player salaries. It is an example of the knapsack problem to figure out how to field a team that stays within this budget cap but still wins games. Other examples include identifying the least wasteful way to cut raw materials, constructing investment portfolios. The knapsack problem also plays a role in some kinds of cryptography.

The knapsack problem has been studied for over 100 years. As the number of objects to be considered grows, the complexity to solve the problem doubles with each additional object to be considered. Practically speaking, even large improvements in computer capacity would be very hard-pressed to keep up with even moderately complex problems with combinatoric properties similar to the knapsack problem. Search problems with a similar structure are extremely unlikely to be solvable without some kind of advanced heuristic to limit the range of alternatives that must be considered. Approximate solutions to the knapsack problem have, in fact, been used for many years.

Despite the success of some forms of computational intelligence, such as expert systems, interest and funding for computational intelligence fell off in the 1970s and early 1980s. Neither the computer resources nor the means of building artificial intelligence were sufficient for dealing with even moderately sized problems. Promises of computers soon being able to do any job that a human could do raised expectations that could not be met. The resulting disappointment led to what has been called an "AI winter."

Ultimately, Turing's claims that intelligence could be implemented by an effective procedure, and that it could be measured by a conversation, along with Newell and Simon's extension of that notion to the physical symbol system hypothesis, turned out to actually interfere with computational intelligence. On the other hand, another of Turing's (1947/1986) ideas turns out to be very useful for the development of computational intelligence, the idea that an infallible computer cannot be intelligent. If we give up on the notion that a computer should be infallible, then we may be able to have a computer yield intelligence. This idea is parallel to the position of Lakatos, discussed in the preceding chapter. Logical positivism sought to create an infallible science and ended up not being able to do science at all. Lakatos and others sacrificed infallibility in favor of criticism and comparative analysis, recognizing the science makes commitments that can be judged, but not proved. In other words, heuristics, which mostly work, but cannot be guaranteed to work, are essential. They may lead sometimes to the wrong answer, but without them, we get no answer at all.

These ideas about fallible intelligence were not widely known in the 1970s and 1980s, even while the failure of the overpromised physical symbol system approach was underdelivering. Interest in and support for computational intelligence in the physical symbol system approach largely disappeared. The introduction of connectionism and other forms of machine learning in the mid-1980s, along with the explosion of the Internet and the World Wide Web, led to a resurgence in interest in computational intelligence, but a different kind of AI, one not so closely tied to the physical symbol system hypothesis. That is the topic of the next chapter.

Definition of General Intelligence

As with studies and tests of human intelligence, formulating a broadly applicable definition of artificial general intelligence remains challenging, and many of the challenges are similar to those involved with human intelligence. But when designing computational intelligence systems, the question is not how to discover what is general, as Spearman tried to do, but to engineer it.

As with intelligence tests, it is relatively easy to formulate specific forms of intelligence as the ability to solve specific kinds of problems. It is easy to evaluate success when the problems are well-formed and have known, specific solutions. In some views, artificial general intelligence is not much more than putting together a sufficient collection of specific problem-solving modules. If we have enough of the right mix of "modules" for solving the right mix of problems, then the problem of general intelligence comes down to selecting the right one of the modules to apply in a novel situation. Or at least that is one line of thinking. We will consider this approach in more depth along with a more detailed consideration of what artificial general intelligence might look like in chapter 12.

For now, it might be sufficient to mention a few characteristics of general intelligence. Current approaches to artificial intelligence work because their designers have figured out how to structure and simplify problems so that existing computers and processes can address them. To have a truly general intelligence, computers will need the capability to define and structure their own problems. Robert Sternberg (for example, 1985), in the context of human intelligence, argues that intelligence consists of three types of adaptive capabilities: analytic, creative, and practical. Current approaches to computational intelligence do very well with the analytic aspects of intelligence. Computers can calculate many times faster than humans can; can hold more variables in memory; and, above all, can behave more systematically than humans. Computers can outperform humans on tasks that depend critically on these capabilities, but they are woefully lacking on the other two.

Computers are limited in the kind of creativity that they can demonstrate. They can demonstrate some apparently creative solutions to certain kinds of problems if that creativity can be achieved through the optimization of parameters. For example, computers can create some kinds of at least pleasant music that may have never been heard before, but the way they do it is by operating on a representation of music created by a computer scientist. The scientist provides the structure of analysis that the computer executes, picking the features of music that are most likely to be important, and providing examples of music from which the computer can extract patterns represented by the provided features. The computer produces the music by adjusting its parameters in order to more closely approximate some combination of the known examples. The music may be novel, but the process by which it is produced is merely an extrapolation of the patterns that it has been trained on.

Practical intelligence is Sternberg's attempt to escape from the intellectual achievement focus of much human intelligence work. Even people who do not receive a formal education can be intelligent. They may lack some analytical capabilities, but they may have other practical capabilities. Practical intelligence includes what is often called "common sense," things that people know that they probably did not require schooling to learn. Commonsense knowledge is focused on facts about an individual's world. Commonsense facts are widely known but seldom explicitly described. This, too, is one of the reasons why common sense is so difficult for computers so far.

A generally intelligent system should, of course, be capable of solving multiple kinds of problems. When it learns to solve new problems, it should expand its capabilities, not replace existing ones. Many current computer systems suffer from "catastrophic forgetting." When they learn to perform a new task, they "forget" how to perform previously learned tasks. Although there is some work on transfer learning, most of the currently available applications can only solve one problem at a time.

A generally intelligent system must also be capable of generalizing from specific problems to more general principles and be able to exploit general principles to construct specific solutions. It must be able to apply learning from one domain to problems in another.

Although not a definition, precisely, of general intelligence, this list of characteristics is a beginning of what we might be looking for in a general computational intelligence agent. An agent with these capabilities would be able to identify problems that need solutions and would have the insight to structure and solve even unstructured problems. Much of the rest of this book is directed at further specifying these features and how they might be implemented.

Conclusion

The symbolic approach was the first pass at attempts to build intelligent machines. It rests on the idea that the processes that make up intelligence can be precisely described. As a result, it focuses on tasks with easy-to-describe steps. This approach was further reinforced by the idea that intelligence is the stuff that well-educated people do, things like proving theorems and playing chess. After McCulloch and Pitts showed that a brain could be a physical symbol system, there seemed to be no reason that both human and machine intelligence could not be structured in the same way. In the next chapter we consider some nonsymbolic approaches to intelligence.

Resources

Church, A. (1936). A note on the Entscheidungsproblem. *Journal of Symbolic Logic, 1,* 40–41.

Colombo, F., & Gerstner, W. (2018). BachProp: Learning to compose music in multiple styles. https://arxiv.org/pdf/1802.05162.pdf

Cooper, S. B. (2004). The incomputable Alan Turing. In J. Delve & J. Paris (Eds.), *Proceedings of the 2004 International Conference on Alan Mathison Turing: A Celebration of His Life and Achievements.* Swindon, UK: BCS Learning & Development. https://arxiv.org/pdf/1206.1706.pdf

Fernández, J. D., & Vico, J. (2013). AI methods in algorithmic composition: A comprehensive survey. *Journal of Artificial Intelligence Research, 48,* 513–582.

Gödel, K. (1931/1992). *On formally undecidable propositions of Principia Mathematica and related systems.* New York, NY: Dover. (Original work published 1931)

Harnad, S. (1990). The symbol grounding problem. *Physica D, 42,* 335–346.

Kellerer, H., Pferschy, U., & Pisinger, D. (2004). *Knapsack problems.* Berlin, Germany: Springer-Verlag.

Lee, D. (2018, January 10). This $16,000 robot uses artificial intelligence to sort and fold laundry. The Verge. https://www.theverge.com/2018/1/10/16865506/laundroid-laundry-folding-machine-foldimate-ces-2018

Lee, D. (2019, January 7). Foldimate's laundry-folding machine actually works now. The Verge. https://www.theverge.com/2019/1/7/18171441/foldimate-laundry-folding-robot-ces-2019

Lenat, D. B., & Feigenbaum, E. A. (1987). On the thresholds of knowledge. In *Proceedings of the Tenth International Joint Conference on Artificial Intelligence, Milan Italy* (pp. 1173–1182). San Francisco, CA: Morgan Kauffmann.

Lindsay, R. K., Buchanan, B. G., Feigenbaum, E. A., & Lederberg, J. (1993). DENDRAL: A case study of the first expert system for scientific hypothesis formation. *Artificial*

Intelligence, 61, 209–261. https://pdfs.semanticscholar.org/68fd/f29cd90a4e7815d1b41ae5ee51f3e78ba038.pdf

McCarthy, J., Minsky, M. L., Rochester, N., & Shannon, C. E. (1955, August 31). *A proposal for the Dartmouth Summer Research Project on Artificial Intelligence.* http://jmc.stanford.edu/articles/dartmouth/dartmouth.pdf

McCulloch, W. S., & Pitts, W. (1943). A logical calculus of the ideas immanent in nervous activity. *Bulletin of Mathematical Biophysics, 5,* 115–133.

Minsky, M. (1968). *Semantic information processing.* Cambridge, MA: MIT Press.

Minsky, M. (1996). In D. Stork (Ed.) *Hal's legacy: 2001's computer as dream and reality.* Cambridge, MA: MIT Press.

Newell, A., & Simon, H. A. (1976), Computer science as empirical inquiry: Symbols and search. *Communications of the ACM, 19*(3), 113–126.

Nilsson, N. J. (2007). The physical symbol system hypothesis: Status and prospects. In M. Lungarella, R. Pfeifer, F. Iida, & J. Bongard (Eds.), *50 years of artificial intelligence* (Lecture Notes in Computer Science, Vol. 4850, pp. 9–17). Berlin, Germany: Springer-Verlag. http://ai.stanford.edu/~nilsson/OnlinePubs-Nils/PublishedPapers/pssh.pdf

Piccinini, G. (2004). The first computational theory of mind and brain: A close look at McCulloch and Pitts's "Logical calculus of ideas immanent in nervous activity." *Synthese, 141,* 175–215. http://www.umsl.edu/~piccininig/First_Computational_Theory_of_Mind_and_Brain.pdf

Quillian, M. R. (1969, August). The Teachable Language Comprehender: A simulation program and theory of language. *Communications of the ACM, 12*(8), 459–476.

Searle, J. (1980). Minds, brains and programs. *Behavioral and Brain Sciences, 3,* 417–457. http://cogprints.org/7150/1/10.1.1.83.5248.pdf

Searle, J. (1984). *Minds, brains, and science.* Cambridge, MA: Harvard University Press.

Simon, H. A., & Newell, A. (1958). Heuristic problem solving: The next advance in operations research. *Operations Research, 6,* 1–10.

Sternberg, R. J. (1985). *Beyond IQ: A triarchic theory of human intelligence.* New York, NY: Cambridge University Press.

Sternberg R. J. (2018). Speculations on the role of successful intelligence in solving contemporary world problems. *Journal of Intelligence, 6*(1), 4. doi:10.3390/jintelligence6010004

Turing, A. M. (1965). On computable numbers with an application to the Entscheidungsproblem. In M. Davis (Ed.), *The undecidable* (pp. 116–154). New York, NY: Raven

Press. (Original work published in *Proceedings of the London Mathematical Society,* Ser. 2, Vol. 42, 1936–7, pp. 230–265; corrections ibid., Vol. 43, 1937, pp. 544–546)

Turing, A. M. (1950). Computing machinery and intelligence. *Mind, 59,* 433–460.

Turing, A. M. (1986). Lecture to the London Mathematical Society on 20 February 1947. In B. E. Carpenter & R. N. Doran (Eds.), *A. M. Turing's ACE Report and other papers.* Cambridge, MA: MIT Press. http://www.vordenker.de/downloads/turing-vorlesung.pdf (Original work published 1947)

4 Computational Intelligence and Machine Learning

Machine learning in some form is the mechanism by which computers expand their capabilities beyond those that were directly programmed in. Artificial general intelligence will need to be able to learn from its experience. Current approaches to machine learning, on the other hand, depend strongly on how human designers structure the problems. In this chapter, we discuss the fundamentals of machine learning and how it is dependent on choices its designers make.

The early rule-based systems, such as General Problem Solver, viewed intelligence as a formal symbol-manipulating problem. If we could describe the processes of intelligence with sufficient precision, then we could design a set of rules by which a computer could simulate them. They then set about demonstrating such intelligence in the context of other formal systems, such as checkers playing and theorem proving. These are activities that could be purely in the "head" of the computer without having to deal with the messy reality of the world.

Limits of Expert Systems

Unlike theorem proving, the goal of an expert system (see chapter 3) is not solely to manipulate symbols but to solve problems in the world, such as identifying the chemistry of a sample from its gas chromatograph, finding oil in an oil field, or diagnosing medical disorders. The symbols in an expert system are grounded in that they relate to specific features of the physical world. The grounding comes from the rules that the knowledge engineers encode into the system, and it derives from the subject matter experts who use similar rules in their daily practice.

Expert systems require reliable evidence and produce verifiable predictions. They still rely on the assumption that the world, or at least the part of it with which they are concerned, can be described with sufficient detail that a set of rules will be sufficient to solve the problems they are intended to address, but they are grounded in that world.

An expert system consists primarily of a knowledge base, which contains facts about the part of the world that the expert system addresses, and a rule base, which contains tools for reasoning about those facts. It reasons with knowledge of the subject matter and with formal representations of the rules. In the context of the rules, the symbols do not stand for any arbitrary facts, but only for certain very specific kinds of facts.

If the world were different, the symbols would be different. For example, the DENDRAL expert system, discussed in chapter 3, suggested possible molecules that could result from measurement of a substance's molecular weight. It addressed this question: what combination of atoms could return a molecular weight of x? Water has a molecular weight of 18 (two atoms of hydrogen contributing 1 each and oxygen contributing 16 units makes 18). Therefore, when a mass spectrometer detects a molecule with a molecular weight of 18, the most likely molecule that could return that weight is water. If hydrogen weighed more, then water would not be among the symbols suggested by an expert system when it finds a molecule with a molecular weight of 18.

Expert systems exploit the fact that computers are general symbol-manipulating devices so that facts, heuristics, and mathematics can all be represented within one of these programs. The reasoning rules were typically organized into what was called an "expert system shell," and they could be reused with a different set of facts. Logic is formal, and it does not care what we are reasoning about, but the facts are specific to certain tasks. The reasoning of an expert system combined knowledge about the subject matter with knowledge about reasoning to produce its results.

The rules in an expert system are conditional "sentences" like:

- A implies B.
- If A is true, then B is also true.

This form of argument is called "modus ponens," and it works independently of what A and B are. When formal arguments are combined with some specialized domain knowledge, then we might get rules like:

- If X is an animal and X lives in the water and X has scales, then X is a fish.

Building out an effective rule set can be quite challenging. Expert systems often required three groups of people: software developers who write the code that runs the expert system (typically the shell), subject matter experts who know about the problem that was to be addressed, and knowledge engineers who know how to translate the subject matter expert's knowledge into rules that the computer could use. Building out an expert system was a custom project and was generally so difficult that only fairly narrow subjects could be practically captured. They never really had the effect of displacing actual human experts, but when they were used, they could sometimes supplement the expert.

MYCIN consisted of about 450 rules and about a thousand medical facts, mostly about meningitis. By today's standards, it was a modestly complex system, but it took years to develop. The goal of MYCIN was to help diagnose patient infections and recommend appropriate therapies.

The insight of expert systems included:

- Knowledge is an essential part of intelligence.
- It is sufficient to focus on a fairly narrow task that can still be baffling for a novice.
- Practical success could be achieved by systematizing and automating the application of knowledge.

These insights were a big shift in the approach to computational intelligence that continues to this day. The data are the most critical part of any current AI project. The projects tend to focus on solving specific problems, such as detecting spam. Although the computational methods are largely different today from what they were during the years of developing DENDRAL and MYCIN, the idea that we can systematize and automate the application of knowledge is still key to current computational intelligence.

Expert systems are an example of what has come to be called Good Old-Fashioned Artificial Intelligence (called, with a bit of pejorative intent, GOFAI by John Haugeland, 1985). The rules were explicit and built "by hand" by the knowledge engineer and the subject matter expert.

For example, a GOFAI algorithm for playing tic-tac-toe (noughts-and-crosses) might be something like this:

- If you or your opponent has two (Xs or Os) in a line (row, diagonal or column), play on the remaining square in that line.
- Otherwise, if there is a square that creates two lines of two, play that square.
- Otherwise, if the middle square is open, play it.
- Otherwise, if your opponent has played a corner, play the opposite corner.
- Otherwise, if there is an empty corner, play it.
- Otherwise, play on any empty square.
- Repeat until no more choices remain.

Tic-tac-toe is simple enough that we can easily list out all of the rules in our system. But as the problems become more complicated, it becomes increasingly challenging to list out all of the rules, because of the "curse of dimensionality." The more variables or dimensions there are, the more ways that they can be combined.

Listing out the rules is extremely brittle as well, meaning that small changes in the problem can cause big issues for the GOFAI system. With a game as simple as tic-tac-toe, we can easily compare a traditional rule-based AI approach with a machine learning one. The rules are not explicitly listed for a machine learning system; rather they are discovered.

Probabilistic Reasoning

Before we dig deeply into machine learning, one more innovation needs to be considered. The computational intelligence projects that came before expert systems had no tolerance for uncertainty. Facts were either true or they were not. In chess, for example, there is no uncertainty about the positions of the pieces on the board. But the world is not so sure as all of that. Expert systems introduced the idea that given some evidence, a rule might not predict something with certainty, but only with probability.

If it swims in the ocean, then it may be a fish.

MYCIN would report the probability that a certain set of facts might indicate a certain kind of infection.

Here is an example rule from a variant of MYCIN, EMYCIN (Buchanan & Duda, 1982):

RULE160

If:

1) The time frame of the patient's headache is acute,
2) The onset of the patient's headache is abrupt, and
3) The headache severity (using a scale of 0 to 4; maximum is 4) * is greater than 3

Then:

1) There is suggestive evidence (.6) that the patient's meningitis is bacterial,
2) There is weakly suggestive evidence (.4) that the patient's meningitis is viral, and
3) There is suggestive evidence (.6) that the patient has blood within the subarachnoid space

Internally, of course, the rules were not expressed in such clear English, but in the computer language LISP.

Extensions of the approach used in MYCIN led to the development of probabilistic reasoning models. These models allowed systems to make inferences with uncertain data and uncertain rules relating facts to inferences about them. For example, the uncertain facts might derive from unreliable sensors, from subjective patient reports, or from other imperfect measurements. When you are in pain, it may be difficult to decide whether your headache is 3 or a 4 on a 4-point scale. Uncertain rules may result from imperfect relationships where, for example, other unmeasured variables may contribute to complex relations.

Among the systems to represent uncertain reasoning are Bayesian networks, which represent facts as nodes in a network and relations among these facts, including predictions, as links with degrees of probability. Dempster-Shafer theory, introduced by Arthur Dempster and expanded by Glenn Shafer, provided for a mathematical theory of evidence and provided a general framework for combining evidence from different sources to yield a degree of belief.

Leslie Valiant's (1984) framework for probably approximately correct learning described how a machine learning system could learn to approximate a function, decision rule, and so on without having to have an explicit theory of the thing it was approximating. Based on the feedback the learning system received during its training, it could learn a rule that was approximately correct, even if it was considerably more difficult to learn an exact rule.

Finally, Hopfield (1982) and Hinton and Sejnowski (1983) introduced another kind of probabilistic learning in the form of a network learning system called a "Boltzmann machine." A Boltzmann machine is a network of symmetrically connected nodes. Each node makes a probabilistic decision about whether to be on or off depending on the inputs it receives from other units.

Although there were several projects that looked at probabilistic reasoning in the face of uncertainty before the mid-1980s, that is when a major surge in interest reshaped the field of computational intelligence research. These and other systems also prepared the way for the emergence of machine learning as a way to replace or enhance the painstakingly constructed rules of expert systems. Finally, they were critical in the development of artificial neural networks, which have been one of the dominant forms of computational intelligence over the last couple of decades.

Machine Learning

The complexity and effort that went into building bespoke expert systems meant that these systems were fundamentally unscalable. Few organizations had the resources or the interest to build them. In those domains where expert systems were able to be successful, the goal was to automate the application of knowledge held by experts. If new knowledge was to be added to the system, it had to be elicited from the domain expert and encoded by the knowledge engineer. If the expert and engineers thought of a situation, then the expert system might have a rule to deal with it. If they did not, and the system encountered a situation for which it had no rule, it was probably out of luck.

The ability to learn is also one of the hallmarks of intelligence. Although the original perceptron (an early neural network described later in this chapter) in the 1950s included learning capabilities, the mid-1980s saw a rapid increase in the capabilities of machines to learn new information and new rules for themselves. These new systems could not only do what they had been specifically programmed to do but they could extend their capabilities to previously unseen events, at least those within a certain range. Rather than provide a machine with an explicit set of rules concerning what to do in each situation, machine learning provides the machine with implicit rules that allow it to learn what to do in situations.

As a step toward a machine learning system that could play tic-tac-toe, we could simplify the rule set to one:

- Of the available moves, choose the one with the highest estimated value.

But now, of course, the question becomes how does the system come to estimate the value of each move? In essence, the system has to learn the estimated probability of winning the game, given each current position (which squares have been marked by X and which by O).

One set of possible values could be:

- +100 points for winning (three in a row filled with your mark)
- +10 for each two-in-a-line (two of your mark with an empty cell)
- +1 for each one-in-a-line (one of your mark with two empty cells)
- +1 for placing your mark adjacent to your opponent's
- +20 for creating a fork (two lines with two of your marks and one empty cell)
- +10 for a block (filling in the empty cell in a line with two of your opponent's marks)
- 0 otherwise)

The best move is to select the cell with the highest score. Like the GOFAI system, this one will tend to win or at least draw the game, but because the system uses the provided scoring unchanged, it still does not involve any actual learning. It is just a more dynamic way of choosing the moves determined by the rules. In this case, the scoring rules determine how the machine will play.

More interesting from a machine learning perspective, we could start with random scores for each possible move and add a method to adjust the score depending on the outcome of the game. If a move leads to a win, then its value is increased by a small amount. If it leads to a loss, then its value is decreased by a small amount. If it leads to a tie, then the choices might be increased by a smaller amount. This process is called "reinforcement learning." Like the psychological behaviorists, the system is rewarded for winning.

Because tic-tac-toe is a formal problem, two computers could play against one another. Paper and pencil are not required to play, only a way to keep track of which squares hold Xs and which hold Os, and to keep track of which moves lead to wins, ties, or losses.

One of these machines playing against the other might initially use the GOFAI method and the other might run the summation method, adjusting its estimate of the value of each move as it leads to winning and losing. At first, the machine learning system would lose games, but eventually it would adjust its choice pattern to at least tie.

Alternatively, both systems may play using the summation method and both may learn along the way. Over a series of games, choices that led to winning or tying would slowly increase their score, and choices that led to losing would decrease their score. The same kind of method could be applied to other games as well. In fact, this method was used in part by AlphaGo learning to play the game of go (discussed more fully in chapter 6).

During reinforcement learning, the system "tries" to maximize some overall level of reward over many learning episodes. When the result is positive, the steps leading to that result are strengthened. When the outcome is negative, the steps leading to that outcome may be punished. From this experience, the system learns a policy or strategy for choosing moves. Early choices in the chain may be more ambiguous than later ones, sometimes leading to a favorable outcome and sometimes to a loss, but over enough games, some strategies will turn out to be more effective than others and will be selected by the learning system.

Varieties of Machine Learning

There are at least three varieties of machine learning. These are characterized by where the system gets its feedback. We have already described reinforcement learning. Two other forms are called "supervised learning" and "unsupervised learning."

Supervised learning is a more direct method of providing feedback. It is commonly used in situations that require the system to categorize items into two or more classes. In supervised learning, an expert labels several training instances for their category. For example, a person may label a group of pictures as containing people and another group as containing cats. The system could then learn how to identify cats versus people by learning the features in these pictures that lead to more accurate classification. This process is called supervised learning because the training examples supervise the system to produce the correct response. At first, the system would probably randomly assign a picture to one class or another. If it assigned the picture to the correct category, then the features of that picture would be more closely assigned to the correct category. For example, a brown patch may be better associated with cat than with people pictures. If the system categorizes the picture incorrectly, then that connection would be weakened.

In unsupervised learning, the system gets its feedback from the data it is working on without any explicit human feedback. It is called unsupervised learning because no human is needed to provide labeled examples. For example, a system may learn to group similar pictures together—called "clustering." If the system is given pictures of human faces and cats, the human pictures may be more similar to one another than they are to the cat pictures, and the cat pictures may be more similar to one another than they are to the human pictures. The feedback is implicit in the way the system is designed to assess the similarity of items and in the rule that groups them by similarity. The supervision, if you will, comes from the way these feedback mechanisms are designed. Eventually the system comes up with groupings that have the highest within-group similarity and the lowest between-groups similarity that it can find.

In both supervised and unsupervised learning, the system is given an objective. For example, the objective may be to maximize accuracy or to maximize within-cluster similarity relative to the similarity of items in different clusters.

Even among supervised and unsupervised learning, there are many machine learning methods, which appear different from one another on the surface, but are actually very similar at a more abstract level. For example, according to Pedro Domingos, every machine learning system involves three key features:

1. A representation of the items to be learned about, their features, and the problem structure
2. An evaluation method used to assess how well the system is working
3. An optimization method to adjust the system to increase its quality as measured by the evaluation method.

According to Domingos,

LEARNING = REPRESENTATION + EVALUATION + OPTIMIZATION

To reduce it further, machine learning involves a representation of the problem it is set to solve as three sets of numbers. One set of numbers represents the inputs that the system receives, one set of numbers represents the outputs that the system produces, and the third set of numbers represents the machine learning model.

When a system is working to classify pictures of cats versus people, for example, the inputs might represent the pictures as a matrix of pixels. Each point (pixel) in the picture could be represented, for example, as a combination of three integers, one for the brightness at that point of red, one for green, and one for blue. An image might contain 200×200 pixels or 40,000 triples.

The output of the system might be just two numbers, one indicating that the picture is a cat and one indicating that the picture is a human.

The learning part of the system is in the third set of numbers—the model. The role of the model is to map the inputs to the outputs so that pictures that actually contain cats produce the output for cats setting its value to 1.0 and set the output for humans to 0.0. Similarly, pictures that actually contain humans set the output for cats to 0.0 and set the output for humans to 1.0. The numbers in the model group reflect the parameters of the model, and those parameters can be very complicated. They are adjusted to produce the correct mapping using some kind of optimization method.

The music recommendation system Pandora represents each song that it catalogs according to about 400 traits, which they call the music's genome. These traits include whether or not the piece involves acoustic rhythm guitars, a repetitive chorus, characteristics of its harmony, rhythm, melody. These characteristics were originally selected by music experts. They are assigned to each song by experts. Every song is represented by some combination of these 400 traits in numeric form.

Pandora uses a machine learning system to identify the characteristics of the music that the person likes and then recommends similar music. Pandora's machine learning system is an example of supervised learning. The listener's choice of songs provides the supervision that allows the system to identify other songs with a high probability of being enjoyed.

The model part of the representation concerns how the system represents the problem of music selection. If it depends on similarity between two pieces of music, then the features that it uses to measure similarity have to be chosen, and the statistical choice has to be made concerning how to compare those features.

Recommendation system models usually represent the similarity of songs to one another and the similarity of users to one another. A complex model is created that recommends songs that are similar to ones that the user has liked before and songs that similar users have liked in the past. The objective is to maximize the probability that a user will like recommended songs. The input is information about each song and each user, and the output is the song being recommended.

To change domains, a chess-playing computer program became a lot easier once it was realized that a chess game can be represented as a tree of moves. Each turn presents a series of potential moves that could be made from that position. Each of the potential moves is a branch on the tree, each of the potential moves from there is a branch from that branch, and so on. Once chess was represented as a tree, the problem became one of selecting a subset of the tree's branches for analysis, because chess trees have too many branches to evaluate them all in a timely manner. The game of go was thought to be unsolvable because the tree contains more branches than the number of estimated particles in the universe. So a critical part of the representation is the set of heuristics that allow the problem to be addressed in an acceptable amount of time.

The representation is chosen by the designer of the system. In many ways, the representation is the most crucial part of designing a machine learning system. So far, no existing computer system can create its own representations yet. Much of the progress in computational intelligence over the last 30 years has been driven by the clever representations, particularly, the heuristics, that computer scientists have invented.

Once a representation has been selected, the next most important choice is an evaluation function. Like the simple tic-tac-toe point system described earlier, an evaluation function determines how far one is from the goal or learning objective. The evaluation method allows the system to choose between alternative moves. Typically, a system prefers a move that gets it closer to its goal.

In the case of tic-tac-toe, the method was to choose the square with the highest expected payoff as represented by the points assigned to each move. In chess, the goal is, of course, to win the game, and the best available move is the one that makes winning most probable. An implicit assumption of most machine learning is that it is possible, at each point in

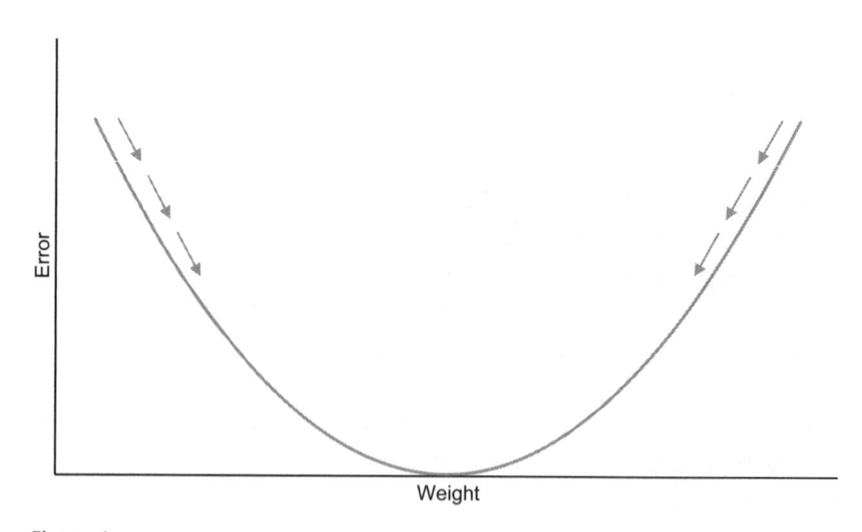

Figure 4
An illustration of gradient descent. Machine learning adjusts parameters to lower the error, that is, the disparity between where the system is and its goal.

time, to evaluate the available choices and to choose the one that will best lead to the goal.

Finally, every machine learning system needs an optimization function. The optimization function is a plan of how to make choices that bring the system closer to its goals. In the tic-tac-toe learning method, the optimization function was to adjust the model weighs based on whether the machine won or lost. Then, during play, it would choose the square that had the points to be earned on that move. The points stand in as a way of estimating the likelihood of winning given each move choice.

There are many different optimization methods that a designer could choose. Some of these functions are better suited to some representations than others. Many of them involve heuristics allowing the system to avoid having to exhaustively evaluate all choices.

One optimization method is based around what can be called "gradient descent" (see figure 4). Every choice that brings the system closer to its goal is a decrease in the distance to the goal, often called "error." So the system can select moves that lower this distance.

The concept of gradient descent is easy to understand when we look at each parameter separately. Each parameter typically starts with a random value. Then, at each step in the learning process, the optimization method adjusts the parameter, either a little higher or a little lower. If the parameter is too high to give the desired result, then it is adjusted lower. If the parameter is too low to give the desired result, then it is adjusted higher. The pattern of adjustments follow the slope (gradient) of the effect the adjustment has on the error, always working to make the error lower and thereby descending the gradient.

All types of machine learning fit within Domingos's description of abstract machine learning. It is an open question at this point whether these kinds of machine learning are all that would be needed for general intelligence. We will return to consideration of this question later in the book. First, let's look at machine learning in more detail.

Perceptrons and the Perceptron Learning Rule

Frank Rosenblatt, in 1957, developed the neural ideas of McCulloch and Pitts into an algorithm called the perceptron. The perceptron was conceived as a kind of neural network, where the neurons were simulated or emulated by circuits or software. The immediate goal of the perceptron work was to take a pattern of inputs, such as a simple picture, and categorize them.

In one early implementation, a perceptron device was constructed with an array of 400 photocells, arranged in a rectangular matrix. The photocells were randomly connected to electronic circuit "neurons." When one of the photocells was activated by a light in that position, it would transmit an electrical signal to all of the neurons to which it was connected. For example, the pattern might consist of a letter, which would illuminate one set of photocells, or another letter that would illuminate a different set.

The system would learn to classify patterns on its photocells by adjusting the weights of the connections, using the perceptron learning rule, similar to the idea of gradient descent described earlier. In some versions of the perceptron, the weights were implemented using variable resistors (potentiometers), which could reduce the voltage transmitted from one neuron to another. In computer simulated versions, the weights were purely mathematical and could be either positive or negative. The higher the weight, the more activation is transmitted. See figure 4.

Each pattern of light and dark presented to the photocells activated a pattern on the simulated neural input. Each illuminated photocell would transmit a 1.0 and each dark photocell would transmit 0.0 to the neurons

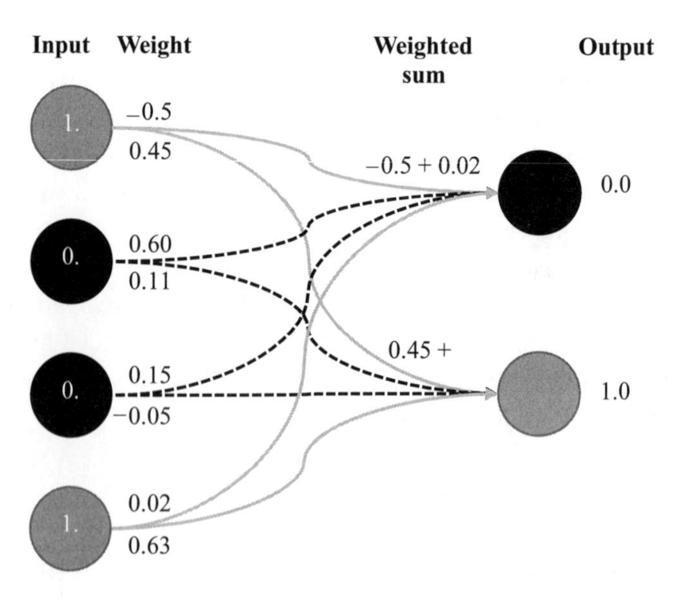

Figure 5
A small example of a perceptron.

to which they were connected. Each pattern was intended to turn on one of the output neurons and turn off the others. The perceptron learning rule specifies how to adjust the weights to achieve this mapping from input pattern to output as the training patterns are presented.

For example, the input might light up the photocells to form the letter H. The desired output from the network would be that the eighth output neuron would have an output of 1.0 and all of the other neurons would have an output of 0.0. A different neuron output, say the first one, would be intended to have an output of 1.0 when the pattern for the letter A is presented.

Figure 5 shows a small example perceptron. The illuminated photocells would provide the input. In this example, the first and fourth units are illuminated; the other two units are dark. The weighted connections transmit their activity to the outputs. The output units sum the weighted inputs that they receive, and if the sum is above a threshold, they respond with a high output; otherwise they respond with a low output. In this example, the second output is activated; the first one is not.

The perceptron learning rule compares the output observed for each given pattern with the desired output. Initially, all of the connection weights

Table 2
The Or problem can easily be learned by a single-layer perceptron.

Input 1	Input 2	Output
0.0	0.0	0.0
0.0	1.0	1.0
1.0	0.0	1.0
1.0	1.0	1.0

are random values, so at the start of training, the perceptron responds randomly to each input pattern. As each pattern is presented to the perceptron, if the network produces an output of 1.0 from a particular neuron when it should be 0.0, then all of the connections from active inputs leading to this output are weakened. If the output is 0.0 when it should be 1.0, then all of its connections from active units are strengthened. After some amount of training, the perceptron weights converge on a pattern that will produce the correct output for each training pattern. The perceptron learning rule is guaranteed to learn any pattern that a perceptron can represent, but not every pattern can be represented by a perceptron.

Table 2 shows an example of a kind of problem that a perceptron could represent. If either input is 1.0, then the output should be 1.0.

The first column shows the first input, the second column shows the second input, and the third column shows the desired output. The first row shows that the two inputs are both 0, and so the desired output is also 0. The other three rows show that at least one input is on (that is, has a value of 1.0), so the desired output is 1.0. Perceptrons can also learn to solve AND problems, where the output should be 1.0 if both inputs are 1.0 and should be 0.0 otherwise.

One of the patterns that a single-layer perceptron cannot learn is called the "XOR" (exclusive OR) problem. The XOR problem involves two inputs and two classes, just like the OR pattern described in table 2. In the XOR problem, however, one class reflects the pattern of either 1, 0 or 0, 1 and the other class involves the input pattern 0, 0 or 1, 1. That is, the output turns on if the inputs are different, but not if they are the same. See table 3.

The first row shows that the two inputs are both 0, and so the desired output is also 0. The second and third rows show that one input is on, the other is off, so the desired output is 1.0. Finally, the fourth row shows that

Table 3
The XOR (exclusive Or) problem cannot be learned by a single-layer perceptron.

Input 1	Input 2	Output
0.0	0.0	0.0
0.0	1.0	1.0
1.0	0.0	1.0
1.0	1.0	0.0

the two inputs are both 1.0, so the desired output is 0. This pattern cannot be learned by a single-layer perceptron. It can only learn problems that are "linearly separable," meaning problems where values above a certain threshold are in one class and values below that threshold are in another. The XOR problem specifies middle values to be in one class and extreme values, both high and low, to be in the other. That pattern is not linearly separable.

The XOR pattern is a fundamental pattern in logic, so a system that is not capable of learning this kind of relationship would have very severe limitations in the kinds of things it could do. Eventually, it was found that a multilayer perceptron could learn this pattern. Stephen Grossberg (1973) described a network where the output of the first group of perceptrons was fed as the input to a second group of perceptrons. He did not know at the time how to train such a multilayer network. When such a rule became widely known around 1986, the use of multilayer perceptrons exploded.

Beginnings of Machine Learning

I want to highlight two features of perceptrons. First, unlike the General Problem Solver or its relatives, the perceptron did not rely on hand-coded rules but learned those rules from examples. These examples included both input and output patterns.

Second, perceptrons employed an optimization process—the perceptron learning rule. At each point in time, the perceptron weights were adjusted to achieve its goal, in this case, to minimize the difference between the desired output and the observed output.

In the perceptron model, the model parameters are the connection weights that transmit activation from the inputs to the outputs. The same model can compute different outcomes by changing its parameters. For example, the simple OR model described above with two inputs and one output has two input parameters: one for each input. If both weights are set to 1.0, then the output would receive activation of 1.0 if either of the inputs is on (1.0) and receive an input of 2.0 if both inputs are on. Any inputs that sum to meet or exceed a threshold value (a third parameter) would turn the output on. Any inputs that sum to less than the threshold would turn the output off. The AND problem could be solved with the same network structure as the OR network, but different weights: input weights of 0.5 and the same threshold of 1.0 for activation of the output. With those weights, both inputs would have to provide 1.0 to get the output to match its threshold of 1.0 (0.5 times the input from input 1 plus 0.5 times the input from input 2).

Machine learning is a combination of statistics and artificial intelligence, leading AI researchers to think in a probabilistic way and to emphasize data over knowledge. Statistical techniques that had been around for as much as a hundred years could be adapted to enable machines to estimate and classify.

Machine learning could be applied to more traditional computational intelligence problems. The Towers of Hanoi problem and the hobbits and orcs problem, for example, were previously represented in terms of specific rules for each move, given the current state. These state transition rules were explicitly written by someone with knowledge about the game and methods for solving it. In these games there could be no unexpected states and the number of potential moves is small enough that a person could write them all down.

This was not the case with checkers, however, let alone with chess. Arthur Samuel (1959) coined the term "machine learning" in the context of describing a program that would learn to play checkers. He chose checkers as a representative kind of problem that could eventually lead to solving more serious problems. Checkers is a relatively simple game, but with enough complexity to be interesting, especially within the limits of computer capabilities of the 1950s.

In this example of machine learning, Samuel used many techniques that would be familiar to modern engineers. He represented the game as a tree, representing all of the legal arrangements that could be reached from any point in the game. He did not have the resources to fully evaluate that tree, so he used heuristics to select which branches were likely to be most successful. He approximated the value of each branch using a scoring function that included things like the number of pieces on the board of each color, the number of kings held by each side, and the number of moves required to elevate a piece to be a king. His heuristic chose the move that would return the highest score under the assumption that the opponent was also choosing moves this way.

The program recorded each position that it had seen in a game along with the ultimate outcome of the game. It then used historical information to augment the move-selection heuristic. He used recorded games played by professionals to further train his system and even had one computer play against another. By 1961, his checker program was able to beat the fourth-ranked checkers player in the United States. By the mid-1970s, his program was good enough to regularly beat respectable players.

Rather than prescribing specific rules to select moves, machine learning provides a mechanism by which the value of each move can be learned. In Samuel's program, the value of memorizing already seen board positions is that his program would have a record of a complete set of branches leading from that configuration. To be sure, that record only included a single path through the tree for each game, but he could know with a high level of confidence what the outcome of that path would be. It effectively substituted data (the results of past experience) for broad reasoning. That, it turns out, is another of the common techniques in modern machine learning. Data, specifically examples of successful problem solving, are much more important for machine learning than is detailed knowledge of the problem state transitions themselves.

The representation, evaluation, and goal combination of a machine learning method describe the means by which state transitions can be learned. For example, a "metarule" in the tic-tac-toe learner, described earlier, specifies that it should choose the square that has the highest value. Another metarule specifies that it should increment the value of a transition when it leads to winning the game.

Another way of saying this is that the representation of a machine learning problem specifies the range of possible state transition operators. The evaluation and optimization methods allow the system to select the appropriate ones. On this view, the learning part of machine learning is just the process of selecting the right values to place on each state transition operator.

Like the weights in a perceptron, parameter values in machine learning are usually not drawn directly from an observation but are estimated from examples. The estimates may not be perfect, but after some training, they are approximately correct. For example, knowing a person's height, we can predict that the person's weight will be within, say, 2 pounds of our prediction or within 5 pounds of our prediction. We can say that our prediction is probably approximately correct. Height and weight are observed, but the relationship between the two of them, the slope of the line relating height to weight, is a parameter we can estimate with the "regression" form of machine learning.

We will assume, in our model, that there is a straight-line relationship between height and weight. In order to estimate weight from height using regression, we must estimate two parameters. One is called the "slope"; the other is called the "intercept." The slope is an estimate of how much weight changes as height changes. If one person is one inch (2.5 cm) taller than another, on average the taller person will weigh more, say, 5 pounds (2.3 kg) more. The slope tells us that for each unit change in height, there will be so much change in weight.

The intercept is the numerically expected weight at zero height. We don't actually expect anyone to have zero height, but 0 is a mathematically convenient and unambiguous value to use to define the position of the line. Once we have an estimate of these two parameters, we fully define the estimated relationship between weight and height. If we measured a new person's height, we could then use this line to estimate what that person's weight would be.

Finding the parameters of a relationship between two variables from examples of people's heights and weights is a standard statistical technique called "regression." No one has to program the estimates of slope and intercept; they are learned from the example data.

Regression is a simple form of machine learning, though until recently it was mostly just thought of as a statistical process. What makes it learning is that its parameters are estimated from example data and are then used to predict values from other data that have not been seen previously (for example, from the heights of never before seen people). The predictions may not be perfect, but they are approximately correct if the system was provided with enough examples.

Similar techniques can be used to make predictions from more complicated combinations of data. Instead of just one estimator (height in this example), complex predictions may involve combinations of many predictors (for example, height, race, gender, zip code, and so forth).

Machine learning is used in many places, from search engines on the web to spam-filtering email, to recommender systems that suggest films, to credit scoring. Machine learning is used to predict categories for items or, as in regression, to predict values from input data, for example, whether a blob in the video camera is due to dust or an obstruction in the road ahead.

Spam filtering is a familiar example of machine learning for categorization that is effective and also relatively easy to understand. A large percentage of the email that many of us receive is unsolicited commercial messages—spam. Like other machine learning tasks, spam filters work from examples—in this case examples of spam emails, the ones you don't want to have to look at, and "ham" emails, the ones you do want to see. One way to get these examples is to ask the user to categorize emails as they arrive. Users can classify emails that they don't want as spam and the ones that they do want as ham.

In one basic form, the spam filter extracts the words or other cues from each of these emails and uses these cues to predict the most appropriate category for it. In this example spam detector, the system counts how often a word occurs in spam emails and how often it occurs in ham emails. It then uses these counts to estimate the probability of the word "Viagra," for example, appearing in spam emails and its probability in ham emails. It repeats this process for every word in each email. Then, when a new email arrives that has not been tagged by the user, the system uses these probabilities to decide whether the new email should be classified as spam or as ham, according to which is more likely to be correct.

Although I have simplified the process a little, this spam filter is an example of what is called a "Bayesian classifier." It is named after an eighteenth-century mathematician and cleric, Thomas Bayes, who described the basic rule on which the classifier is based. Bayesian classifiers are one kind of machine learning that turns out to be particularly effective at separating spam from ham emails. We will discuss Bayesian learning again in chapter 10.

A Bayesian classifier spam filter uses the distribution of words in the two classes to infer whether the set of words in a particular email is more likely to have come from a spam email or a ham email. It learns these distributions from the emails that were labeled by the user. The probabilities of each word for each category are the parameters that the system has to learn. So this form of machine learning involves many parameters.

Learning to filter suspicious emails is an example of supervised learning because it uses the labels provided by a "supervisor," the mailbox owner, to learn to reproduce the decision patterns of that user.

Bayesian classifiers fit neatly into the Domingos framework for machine learning. As noted above, the key features of this framework are representation, evaluation, and optimization.

Representation A Bayesian spam classifier represents the emails it is judging as a "bag of words." The system uses the words and their frequencies, but not their order. It is as if we took all of the words in an email and threw them into a bag. It represents the presence of each word in an email as an array or list of numbers. Each word has a position on the list, and that position is set to 1.0 if the word appears in the email and is 0.0 otherwise.

Evaluation Evaluation of the Bayesian spam classifier consists of the measuring of the probability of a correct classification on these examples. How often did it correctly identify junk emails as spam and good emails as ham?

Optimization Bayesian classifiers have very simple optimization. As we accumulate more examples of spam and ham, we get better estimates of the probability of each word appearing in spam or in ham. We also need to estimate a threshold for deciding whether to classify an email as spam. Emails with probabilities higher than the threshold are called spam, and others are called ham.

How we represent the problem to be solved determines the range of possible solutions. Each combination of potential parameter values is a hypothesis for how to solve the problem. Evaluation and optimization allow the system to select among these hypotheses, but a system cannot select a solution that is not among the set of potential solutions. It cannot just decide that today, it will treat the problem as a regression.

If a problem is represented as a tree, then the only solution that can be reached is one that can be described as following the branches of a tree. Machine learning can be used to select a path through the tree. But at least so far, machine learning cannot tell you that a tree is the wrong representation and suggest the right one.

The optimization method is the means by which the system searches among the potential solutions to find the one that is "best" or as close to best as it can find. There are many optimization methods.

Returning to the space metaphor, the optimization method moves the system closer its goal. The proximity to the goal is measured by the evaluation process.

If there are only a few parameters to be adjusted, then we can use brute force and try them all. But when there are many parameters, the number of ways that they can be combined means that there are too many combinations to consider. Then we need heuristics to select the combinations to evaluate. Improvements in machine learning have often relied on finding better methods to predict which changes will be useful and focusing on those.

Classification machine learning, as discussed, divides the patterns into categories that are determined by the labels assigned to examples. The system is expected to generalize from these examples to previously unseen items. The user/supervisor tells the system how to organize the examples into categories, and the system learns to reproduce this organization.

Besides clustering, another unsupervised machine learning approach is called "association rule learning." For example, in market basket analysis, all of the items that each buyer selects during a trip to a supermarket are tracked. It may be the case, for instance, that people who buy potatoes and onions are more likely than the average shopper to also buy hamburger. If this kind of rule is consistent, then the information could be used to improve marketing.

There is a story that was making the rounds a few years ago about how the retail company Target analyzes their shoppers' buying patterns. According to this story, Target found that women buying supplements like calcium, magnesium, and zinc were also buying higher amounts of unscented lotion. They first observed this in conjunction with women who had signed up for a baby shower registry, indicating that they were pregnant and when they expected to deliver. Target could then use this information to market other baby-related products to these women. Machine learning was used to find the relationships, and Target made use of their discoveries. This story may not actually be true (Piatetsky, 2014), but it illustrates the idea of association rule learning using machine learning.

Machine learning is highly dependent on the examples that it is trained on. As Domingos noted, success depends on good data even more than on having good algorithms. What is worse is that poor data can deceive users into thinking that they have an effective machine learning process when really the system has merely learned some artifact of the data. For instance, if one were training a neural network machine learning system to recognize pictures with cats in them, what the system would learn would depend on the distractor pictures (the pictures of noncats) that were also presented. If the alternative category contained only landscapes, for example, the system might learn to recognize the difference between pictures with large amounts of green and blue versus pictures with large amounts of black, brown, orange, or gray. The system would appropriately distinguish between cats and landscapes in the context of those specific pictures, but learning to categorize pictures by color is far different from learning to categorize them by whether they contain certain objects.

People looking at the pictures might jump to the conclusion that the system learned to identify cats, but that does not mean that the category "cats" is really what it learned. In fact, a recent analysis by Raghavendra Kotikalapudi found that one network was actually learning to categorize the objects in pictures by often trivial properties of those objects. For example, the feature that it used to identify pictures of penguins was the large white area of its belly. The machine learning system is constrained by the representation designed into it, but it can still learn surprising things about the properties of that representation, things that do not necessarily match the expectations of its designers.

A related problem is called "algorithmic bias." A machine learning system may be perceived as objective, and in the context of the data on which it was trained, it really is, but the system cannot be any more objective than the data on which it was trained. For example, in 2015, Google's picture classification software was called out for falsely classifying a picture of a 21-year-old African American programmer as a gorilla. Chimp also matched. The point is not that the software was foolish—maybe it was—but more that the distinction that the computer learns is not always the distinction that its designer wants it to learn. Better data, more labeled examples of black people, for example, might have prevented this problem.

The particular examples used to train a machine learning system are selected in some way, and this selection influences the outcome of the machine learning model. The system designer also selects the features to be used as the elements of the machine learning model. For example, machine learning has been used to create models of recidivism, the likelihood that a criminal will commit another crime in the future. Many of these models are highly controversial for a number of reasons, but I want to think about two of those reasons here; both involve the perceived fairness of the system's recommendations.

Legally, each person is entitled to be treated fairly by the government, including the courts, as an individual. The person should be judged on his or her own merits. When the output of the recidivism program was analyzed, however, the investigators found that it treated black defendants differently from white defendants.

When the system made a mistake on a white defendant, it erred by predicting that the white person would not commit an additional crime when the person actually eventually did. When the system made a mistake on a black defendant, it erred by predicting that the person would commit a crime when the person did not. Overall, the system was moderately accurate, but the difference in the kinds of errors made with white versus black defendants was seen to be unfair.

The solution to this bias is to include fairness in the definition of the system's goal. Currently most machine learning is designed to maximize the correctness of its predictions, but it could easily be designed to maximize both the correctness and the fairness of its predictions. If a system does not include a definition of fairness in its goals, then any achievement of fairness will be merely accidental. Fairness is too important to be left to the whims of accident.

Reinforcement Learning

Reinforcement learning was introduced earlier in the context of learning to play tic-tac-toe. As a form of machine learning, it is in between supervised and unsupervised learning, because individual examples are not labeled, but the system still gets feedback when it reaches or fails to reach its goals.

The learner is not told which actions to take but must form a policy to select an effective series of actions that will lead it eventually to its desired outcome (and the reinforcement).

The goal of a reinforcement learning system is to maximize some cumulative "reward." A major problem is that the ultimate reward may depend on a series of actions and so be delayed for some time after those actions have been performed. Contrast reinforcement learning with supervised machine learning, where the feedback immediately follows each action. The challenge for the optimization method in reinforcement learning is to determine how to allocate the credit for the observed outcome to the actions that ultimately led to the reward.

In general, reinforcement learning and credit assignment (that is, how to allocate some proportion of the ultimate reward to earlier choices) are intractable problems to solve perfectly. Getting to rewards can require a large number of steps chosen from an even larger set of possible actions. It is impossible to check, in anything but trivial systems, all possible combinations of all possible actions that can lead to reward. As with other machine learning situations, heuristics are needed to simplify the problem into something more tractable, even if it cannot be guaranteed to produce the best possible answer.

There may be several actions to choose from at any point in time, but the only available information is the machine's current state, which includes its history of actions and past reinforcements.

A reinforcement learning system needs to learn the probabilities that each action will lead to the reinforced outcome. The relation between each individual action and the ultimate receipt of reinforcement may be tenuous. Actions may not always succeed. So the system's learning methods must be able to cope with errors and with behaviors that individually affect the likelihood of reward by only a negligible amount.

Examples of reinforcement learning include robots learning to navigate their space and stock market investing. Picking a stock at a particular time does not immediately yield a profit—the reinforcement. Rather, the profit appears only some time later when the stock has been sold. Selecting which crops to plant and when to plant them can also be addressed with reinforcement learning.

Reinforcement learning is particularly suited to interactive problems, where one can only learn about a situation by interacting with it. In these cases, it is difficult to get examples of desired behavior that are correct and are representative of the kinds of situations that the agent may encounter.

Summary: A Few Examples of Machine Learning Systems

The variety of tasks to which machine learning has come to be applied is growing rapidly. In many cases, successfully solving a problem using machine learning causes the machine learning to effectively disappear from the user's sight. Speech recognition, for example, has succeeded largely because of effective machine learning to map speech sounds to text, but following this success, people do not generally still think of speech recognition as a machine learning problem.

Credit card fraud

Machine learning has been used for several years to categorize credit card transactions as either potentially fraudulent or genuine. Transactions that are potentially fraudulent are routed for further processing and follow-up. They may be stopped at the point of sale.

Product recommendation

When you buy something on Amazon or from other online retailers, they will often offer you other products that you might like. Similarly, Netflix recommends movies. These recommendations come from recommender systems. There are several different ways in which the machine could learn what you might like. For example, the system may have learned that people who buy shampoo also tend to buy deodorant. If you buy shampoo, it will then recommend deodorant.

Face recognition

Applications such as Facebook analyze photographs to identify the people depicted in them. Face recognition combines several machine learning problems, first to identify that there is a face in a picture, and then to identify whose face that is. Both kinds of recognition may depend on identifying face parts (such as noses or eyes) or other features of faces.

People find recognizing faces easy. However, computer face recognition is a challenging problem because of the large number of people whose faces could appear in a picture and because of the complexity of face geometry.

Machine learning for face recognition involves a series of steps, each of which is a machine learning problem on its own, each involving a substantial number of training examples. When put together, however, it results in an effective method to identify the individuals in pictures in a wide variety of positions and against a broad set of backgrounds. But it bears little resemblance to the process apparently used by people, so when it makes a mistake, it is likely to make a different kind of mistake than a person would make.

Conclusion

Machine learning is an area of very active research. Machine learning is a critical part of computational intelligence and one whose importance continues to grow. Using a variety of techniques and clever representations, it allows machine to accomplish tasks that they were not programmed to accomplish in response to objects that they may have never seen before. The main insight that led to machine learning was the combination of statistics with probabilistic learning.

Resources

Buchanan, B. G., & Duda, R. O. (1982). Principles of rule-based expert systems. In M. Yovits (Ed.), *Advances in computers* (Vol. 22) (pp.163–216). New York, NY: Academic Press.

Domingos, P. (2012). A few useful things to know about machine learning. *Communications of the ACM, 55*(10), 78–87. doi:10.1145/2347736.2347755; http://homes.cs.washington.edu/~pedrod/papers/cacm12.pdf

Dressel, J., & Farid, H. (2018, January 17). The accuracy, fairness, and limits of predicting recidivism. *Science Advances, 4,* eaao5580. http://advances.sciencemag.org/content/4/1/eaao5580

Duhig, C. (2012). How companies learn your secrets. *The New York Times Magazine.* http://www.nytimes.com/2012/02/19/magazine/shopping-habits.html

Geitgey, A. (2016). Machine learning is fun! Part 4: Modern face recognition with deep learning. https://medium.com/@ageitgey/machine-learning-is-fun-part-4-modern-face-recognition-with-deep-learning-c3cffc121d78#.kleeml9o7

Grossberg, S. (1973). Contour enhancement, short term memory, and constancies in reverberating neural networks. *Studies in Applied Mathematics, 52,* 217–257.

Haugeland, J. (1985). *Artificial intelligence: The very idea.* Cambridge, MA: MIT Press.

Hinton, G. E., & Sejnowski, T. J. (1983, May). Analyzing cooperative computation. In *Proceedings of the Fifth Annual Congress of the Cognitive Science Society*. Rochester, NY.

Hopfield, J. J. (1982). Neural networks and physical systems with emergent collective computational abilities. *Proceedings of the National Academy of Sciences of the United States of America, 79,* 2554–2558.

Howley, D. (2015, June 29). Google Photos mislabels 2 black Americans as gorillas. Yahoo Tech. https://www.yahoo.com/tech/google-photos-mislabels-two-black-americans-as-122793782784.html

Koerth-Baker, M. (2016). The calculus of criminal risk: The justice system has come to rely heavily on quantitative assessments of criminal risk. How well they work is a complicated question. http://undark.org/article/of-algorithms-and-criminal-risk-a-critical-review

Kotikalapudi, R., & Contributors (2017). Keras-vis. Github. https://github.com/raghakot/keras-vis

Lindsay, R. K., Buchanan, B. G., & Feigenbaum, E. A. (1993). DENDRAL: A case study of the first expert system for scientific hypothesis formation. *Artificial Intelligence, 61,* 209–261. doi:10.1016/0004-3702(93)90068-M; https://profiles.nlm.nih.gov/ps/access/BBABOM.pdf; https://stacks.stanford.edu/file/druid:jn714xp6790/jn714xp6790.pdf

Piatetsky, G. (2014). Did Target really predict a teen's pregnancy? The inside story. http://www.kdnuggets.com/2014/05/target-predict-teen-pregnancy-inside-story.html

Samuel, A. L. (1959). Some studies in machine learning using the game of checkers. IBM *Journal of Research and Development, 3,* 210–229.

Shafer, G. (1976). *A mathematical theory of evidence.* Princeton, NJ: Princeton University Press.

Shafer, G. (1985, July). Probability judgment in artificial intelligence. In L. Kanal & J. Lemmer (Eds.), *Proceedings of the Workshop on Uncertainty and Probability in Artificial Intelligence* (pp. 91–98). Corvallis, OR: AUAI Press. https://arxiv.org/ftp/arxiv/papers/1304/1304.3429.pdf

Valiant, L. G. (1984). A theory of the learnable. *Communications of the ACM, 27*(11), 1134–1142.

Valiant, L. G. (2013). *Probably approximately correct: Nature's algorithms for learning and prospering in a complex world.* New York, NY: Basic Books.

5 Neural Network Approach to Artificial Intelligence

This chapter continues our discussion of machine learning, extending it to simulated neural networks. These systems, starting with the perceptrons discussed in the previous chapter, are inspired by the operations of neurons, the computational elements of the brain.

Human intelligence is a product of mental operations, which are implemented by the neurons of the human brain. The perceptron, described in chapter 4, was an attempt to use brain-like processes to implement a machine learning system. There was speculation that if brains can implement intelligence, then maybe we can achieve similar results by emulating the computational methods of the brain. Progress in neuroscience, an emerging understanding of how the brain's neurons and basic networks of those neurons work, further lent enthusiasm to this approach. But this growing neuroscience knowledge also recommends caution in our attempts to model the human brain. As much as we know about neuroscience, we are still very far from having a complete understanding of how the brain actually implements intelligence. Nevertheless, simulated neurons turn out to provide an extremely powerful model for machine learning.

Networks of simulated neurons, the simplest of which is arguably the perceptron, have been able to solve a number of machine learning problems that were intractable with other forms of computational intelligence. These neural networks consist of usually layered collections of simulated neurons where each layer but the first receives inputs from the neurons in the previous layer and each layer but the last provides inputs to the succeeding layer. These simulated neurons are an abstraction and a simplification of actual biological neural networks.

Biological neurons consist of three main parts, a cell body, an axon, and dendrites. The cell body controls the biological activities of the cell. The axon is the fiber that transmits the messages from the neuron to the dendrites of other neurons. The dendrites are a treelike structure, often with many branches, that receives information from the axons of other neurons. A given neuron can connect to thousands of other neurons.

Neurons do not connect directly to one another. Rather, there is a tiny gap, called a synapse, over which small packets of chemical signals are "sent." The axon of the transmitting neuron releases a packet of neurotransmitter, which then passively diffuses across the synaptic gap and binds with receptors on the dendrites of the receiving neuron. The process of binding the neurotransmitter to the dendrite causes a cascade of chemical reactions in the receiving cell.

Simplifying a great deal, when neurons receive signals from other neurons, they add up the activation they receive. Some of the transmitting neurons contribute excitatory signals to the receiving neuron, and some contribute inhibitory signals. Each type of signal is mediated by different neurotransmitters. If a neuron receives enough excitation, it will become active, resulting in a spike of electrical activity and releasing its own neurotransmitters from its axons. The spike can be measured electrically because the activity of the neuron causes an electrical current to flow through the cell membrane.

Computational neurons generally do not represent the details of spiking, but rather model the overall activation of the neuron. They sum up the inputs they receive from other simulated neurons, some of which have positive weights and contribute to activation and some of which have negative weights and subtract from activation. The output of a simulated neuron is typically 1.0 if the total of inhibition and excitation exceeds some threshold and is typically 0.0 (or, in some networks, –1.0) otherwise. Other simulated neurons output a value that is related to the sum of activity they receive. At low sums, the output is effectively 0.0; at high sums, the output is effectively 1.0; and in between, the activity is approximately proportional to the sum of the inputs.

Another simplification of computational neurons is that their role is fixed in the networks they constitute. In contrast, some recent experiments have found that biological neurons may change their role over time. Neurons that at one time reliably signaled that a mouse would make one response later signaled the opposite response when the activities of these neurons were recorded over the course of weeks.

The simplified model of computational neurons used in neural network models cause each neuron to change its activity pattern over the course of training to optimize its role in producing the desired outputs for each input. Studies like this one with the mouse suggest, instead, that the brain changes dynamically and the role played by individual neurons may change over relatively short time spans. We do not yet know what the implications of this dynamic change are for intelligence, but it could indicate that we still understand very little about how neurons mediate intelligence in the brain.

There were three big problems with the simulated neural approach at the time of the initial work on the perceptron. One was the general lack of computing power—an issue that continues to limit computational intelligence to this day. Without sufficient computational power, a neural emulation might, in fact, be able to demonstrate intelligence eventually, but the designers would all have died before a problem was solved.

The second major problem with the perceptron approach was an overpromise on its capabilities. There were severe limitations in the kind of logical functions a perceptron could compute. As discussed in chapter 4, for example, a perceptron cannot learn an exclusive OR (XOR) pattern that responds positively if one of two inputs is active and negatively otherwise. Without these patterns, the perceptron could not function as a complete model of logic, let alone as the basis for intelligence. Although not published until 1969, Marvin Minsky and Seymour Papert's critique of the perceptron had a devastating effect on perceptron and other neural network systems. By the 1970s, there were only a few pockets of neural network research actively working.

The third problem with the initial approach to neural networks as the basis for intelligence was a lack of a learning rule to allow a network to learn how to compute the mapping from the inputs to the outputs, except in single-layer systems.

McCulloch and Pitts had shown how networks of neurons might implement all of the necessary properties of a Turing machine, but they did not have a way, other than laying out the network by hand, to generate structures with these properties. They did not know of a method by which the neurons could organize themselves, though obviously they do in the human brain.

The perceptron learning rule was helpful, but it could only be applied to a single-layer network. Minsky and Papert (1972, p. 32) argued that the same limitations would apply to multilayer networks if we had a way of training them. As it turns out, they were wrong.

Although it was not well-known at the time, Belmont Farley and Wesley Clark, in 1954, had already come up with a method for training two-layer networks and had simulated them on an early digital computer. Farley and Clark's networks, which contained up to 128 neurons, were trained to recognize simple patterns. They used a learning rule similar to what later became the perceptron learning rule.

The situation with self-organizing networks changed dramatically in 1986 with the publication of an article on backpropagation in *Nature*, by David Rumelhart, Geoffrey Hinton, and Ronald Williams, and a two-volume set of technical books, edited by David Rumelhart, James McClelland, and the PDP Research Group. These books described what they called "parallel distributed processing" (PDP)—neural networks and related structures. The largest impact arguably came from their description of a "backpropagation" learning algorithm that could be used for multilayer networks. The basic idea of backpropagation is that each neuron would have its weights adjusted according to that neuron's contribution to the overall network error.

Their method was an extension of the perceptron learning rule similar to that described by Rosenblatt, Farley and Clark, and others. Their specific method was described earlier by Paul Werbos in his 1974 dissertation, but, again, it went largely unnoticed until it was popularized by the PDP books. Backpropagation could allow a multilayer perceptron to learn to set its weights. The kinds of problems that left perceptrons baffled were now readily addressable by perceptron networks that fed into perceptron networks—multilayer perceptrons trained with backpropagation.

The PDP books had a profound effect on artificial intelligence and on cognitive science in general. Interest in the PDP models exploded. It is difficult to overstate the excitement that these books engendered.

Neural Network Basics

The basic idea of neural networks is that information and processing are represented by patterns of activation across a network of neuron-like units and the connections between these units. Each unit is active to varying degrees and can transmit this activation to other units to which it is connected. Each unit receives activation from the neurons feeding it and transmits activation to the units it feeds.

In a multilayer perceptron, the units are organized into layers. The initial layer receives inputs from the environment, such as a pattern or light (on) or dark (off). The units of the input layer then feed units in a second layer (often called the "hidden layer"), and the hidden layer feeds an output layer. In a multilayer perceptron, the units in a layer are connected only to units in the subsequent layer, not to each other. A given unit may receive inputs from multiple units in the previous layer and may send outputs to multiple units in the next layer.

Each unit has a threshold for activation. If the sum of the inputs it receives is below the threshold, then the unit will be off, or inactive. If the received sum is above the threshold, then the unit will turn on, or be active. Learning consists of adjustments of the connection weights between one neuron and the next. Higher weights contribute more to the sum at the receiving unit; lower weights contribute less, or even negative amounts.

Knowledge in the network is represented by the strength and pattern of connections among these units. Unlike expert systems, knowledge and processing of that knowledge are inseparable in a neural network.

The perceptron learning rule showed how to adjust the connection weights from the inputs to the outputs of a single layer to compensate for errors in the output of the network. For example, the goal of the network may be to distinguish male from female pictures. If the current input represented a "male" pattern, then the desired output would be for output neuron representing a "male" decision to be active and the other neuron to be inactive. If the current input represented a "female" pattern, then the desired output would be the opposite. Recall that the perceptron learning rule would adjust the input weights of these units to more closely approximate the desired pattern in just this way.

The backpropagation learning rule extends a similar learning method to multilayer networks. The networks are trained on a set of labeled instances, where the label specifies the desired output for each input. Each connection weight is adjusted by the backpropagation learning rule proportional to that neuron's contribution to the error. The amount of error is said to propagate from the output layer through preceding layers, adjusting connection weights as it goes. If a connection to a hidden layer unit was strong, and that hidden layer unit contributed to the output error, then the strong weight would be weakened.

Whether the network consisted of a single layer or multiple layers, the inputs to the neurons could be described as a vector, which is just an ordered list of numbers. Recall that one of the earliest perceptrons had 400 photocells as inputs. The amount of light projected onto each of these 400 photocells could be represented as a number (say, between 0 and 255). If we numbered the photocells from 1 to 400, then the light received by each of them could be written in a list where the first number in the list represented the input to photocell 1, the next number represented the input to photocell 2, and so on. That would be a vector.

The output of the network can also be described as a vector. Sometimes only one output neuron is desired to be on (have a strongly positive output value) and all the others are intended to be off. If the purpose of the network is to identify the uppercase alphabetic character projected on the photocells, then we might have 26 outputs, one for each letter. If the purpose is to distinguish cat pictures from other pictures, then we might have just two outputs. When a cat picture is presented, the desired output vector might be [1.0, 0.0], and when a person picture is presented, the desired output might be [0.0, 1.0].

Neural networks succeed because they do not base their operation on categorical rules but on pattern recognition. They recognize that a fever of 99.9 degrees is almost as dangerous as a fever of 100 degrees. They can reason probabilistically, for example, increase the likelihood of saying that someone should be treated with aspirin the higher the person's temperature is.

The metaphor of neural network approaches to artificial intelligence is to solve problems in a brain-like way, rather than in a computerlike or Turing-machine-like way. Rather than being based on categorical rules that either apply or do not apply, as is the case with expert systems, neural networks treat inputs as more or less similar to one another, using units that may be more or less active. Some people call this style of computation subsymbolic because the information is represented in the activation of units rather than in symbols written on a virtual tape.

Neural networks make it possible to solve problems that have been resistant to physical symbol system approaches, including expert systems.

Symbols are all-or-none. In a Turing machine or its equivalent, a symbol is either written or it is not. A rule is either applied or it is not. In contrast, in a neural network, a representation can occupy a state that is somewhere between being fully present and being fully absent.

Neural networks specialize in fuzzy representations. Symbol systems specialize in "crisp" representations. With some effort, each one can emulate the other. Digital computers can simulate fuzzy neurons, and fuzzy neurons can simulate discrete categories. For example, CD, DVD, and Bluray disks or downloaded audio or video programs are digital copies of an analog performance. Human brains can think logically and speak symbolically. McCulloch and Pitts showed how a network of neurons could implement a Turing machine. But the fundamental building blocks of these two approaches are distinctly different.

Neural networks are much more about patterns than they are about symbols and rules. Resemblance is a key relation. Face recognition, for example, is much more about identifying faces that resemble those that we know about. The overall pattern of similarity is important; no one feature or even no small set of features may define this relationship. When combined with efficient machine learning methods, neural networks can consider many thousands of variables in their computations, with each of these contributing a small amount but all of these together forming a pattern. The number of factors that can be considered by a neural network can far overwhelm any knowledge engineer trying to formulate comparable rules.

Dolphin Biosonar: An Example

My colleagues and I used neural networks, for example, to model how dolphins use their biosonar to recognize objects underwater or even under mud. Dolphins, like bats, use sound as an active source of information about their environment and the things in it. Both kinds of animals can see when there is enough light, but bats flying in dark caves and at night and dolphins swimming in deep water may not have enough light to see effectively. Dolphin vision, in particular, is like that of many terrestrial mammals, but dolphins are really effective at using sound waves to get information about their underwater world.

In experiments, we found that a dolphin can tell the difference between two cylinders that differ in thickness by the width of a human hair at an underwater distance of 8 meters (about 26 feet). Unless the water is very clear, people cannot even see the cylinders at that distance, let alone tell them apart.

Dolphins using their biosonar send out a very brief click, which they generate inside their head. The click enters the water through the melon, that bulbous structure on the front of a dolphin's head. The click is about 50 microseconds (50 millionths of a second) in duration. It travels through the water in a tight beam until it reflects from the object, making an echo. The dolphin picks up the echo through fat channels in its jaw, which transmit the sound to the dolphin's inner ear.

The pitch of a sound, measured in Hertz (abbreviated as Hz or kHz for kilohertz; one kilohertz is 1000 Hertz), corresponds to the sound's frequency, like the keys on a piano. Piano keys on the left side of the keyboard are low notes and correspond to low frequencies (around 27.5 Hz). Keys on the right side of the keyboard are high frequency notes (up to about 4 kHz). People can hear sounds up to about 20 kHz, at least when they are young. Dolphin hearing, on the other hand, extends all the way up to 150 kHz. We humans need special equipment just to detect frequencies that high, but dolphins hear them naturally.

When an echo of a dolphin click reflects off an object, it contains a mixture of many different frequencies. According to our studies and others, the dolphin uses the pattern of that mixture to identify what the object is and many of its properties.

Mammalian ears, including those of dolphins, mechanically transform the frequency pattern of sound into a spatial pattern on the cochlea, contained in the inner ear. The spatial pattern results in a neural pattern that is transmitted to the dolphin's brain.

In our studies, we used a technique called "fast Fourier transform" to analogously measure the mixture of frequencies in a signal and obtain its "spectrum." The spectrum created mathematically more or less what the dolphin's cochlea did biomechanically. A spectrum is a measure of the amplitude of a signal across frequencies.

We trained a dolphin to float in the water with her head in a hoop underwater. We then placed a special underwater microphone (called a "hydrophone") next to the dolphin and recorded the echoes that came back from one of three objects. The dolphin sent out its sonar signals, presumably listened to the echoes that came back, and then touched one of three "target poles" to tell us which object she had been presented with. If the dolphin got it right, we gave her a fish.

We did the same kind of experiment with these targets buried in mud. The dolphin was able to distinguish among cylinders made of coral, hollow aluminum, and foam-filled aluminum, all with high accuracy.

We then repeated the experiment using a three-layer neural network in place of the dolphin. We found that the accuracy of the network was comparable to that of the dolphin, both typically in a range well above 90%.

Pattern recognition, as expected, was critical in the performance of the neural network and of the dolphin. We did many follow-up experiments and found that we could eliminate some frequencies from the spectrum and still get good accuracy. The precise frequencies we eliminated did not matter, but the more we degraded the echo, the lower was the performance of the dolphin and of our network. Apparently, it is the overall pattern of the echo that is important, not any specific features of the echo.

My colleagues and I built this kind of network into a kind of underwater robot which could travel around using its sonar on the ocean floor and identify objects buried in the mud in front of it. There is a video of this vehicle on YouTube (https://www.youtube.com/watch?v=fP9k0eLP4ws).

Interest in neural networks diminished some in the 1990s and early 2000s. Like a lot of interesting tools, they were good for many things but failed to live up to their hype. They still played significant roles in things like noise-cancelling headphones and some cameras. They remained an important part of credit scoring as well, among other applications.

The use of neural networks was limited by scaling issues and by the power of the backpropagation learning algorithm to train networks with more than one hidden layer. In theory, a network with one hidden layer should be as good as one with multiple hidden layers, but multilayer networks may be more practical to build or train. Multi-layer networks may require fewer simulated neurons, and more critically, may require fewer connections, to match a network of equal power with a single hidden layer. A multi-layer network may also require fewer training examples and less time than a single-hidden layer.

Progress came, as it frequently does in artificial intelligence, from improved computational capacity. But two other developments also contributed to the reemergence of neural networks. One of these factors was the availability of so-called big data, ranging from credit card use databases to Google queries. These systems provided large amounts of labeled data that could be used to train large-scale neural networks. These data provided the training examples that were otherwise so expensive to generate (because they took human effort to provide the labels). In these big data sets, the labeling came from the natural consequences of the transactions that were being recorded. The second source of progress was the development of new neural network architectures and stronger methods for training them. For example, Yann LeCun developed *convolutional neural networks* to recognize handwritten digits. Inspired by the receptive field patterns of biological visual processing, the neurons in a convolutional neural network have overlapping receptive fields. This style of network and some relatives have become very important in computational intelligence.

Convolutional neural networks and others, such as recurrent neural networks, formed a new class of neural network, deep learning networks, which may involve many more than three layers. Rather than strictly feeding from one layer to the next, deep learning networks can include more complex connections within a layer. Moreover, each layer could be trained using different learning rules. Some could be trained by supervised learning, for example, using backpropagation, and others using unsupervised learning.

The idea of using different training regimens on different layers is, arguably, the biggest insight in deep learning. Like more traditional neural networks, and like much of machine intelligence, the real genius comes from how the system is designed, not from any autonomous intelligence of its own. Clever representations, including clever architecture, make clever machine intelligence.

Deep learning networks are often described as learning their own representations, but this is incorrect. The structure of the network determines what representations it can derive from its inputs. How it represents inputs and how it represents the problem-solving process are just as determined for a deep learning network as for any other machine learning system.

Every network with a hidden layer, that is, one that is neither the input nor the output layer, learns a representation inherent in the pattern of activation across the input layer. This pattern of representation may have no obvious relationship to the input pattern or nameable features of the input patterns. Deep learning networks may use different learning rules to form those hidden layer representations than a multilayer perceptron would, but the hidden layer representations are still just transformations of the original inputs, not new kinds of representations. The network can select among the kinds of patterns that are available, but it cannot construct new kinds of patterns, and just what it selects is determined by the structure of the hidden layer as provided by its designer.

For example, recall that Quoc Le and his colleagues (2012) built a nine-layer neural network involving a billion connections, on a cluster of 1,000 machines, using 16,000 computational cores. They trained it for three days on a set of 10 million 200 × 200-pixel images. They then examined some of the simulated neurons in this system, and, despite the fact that the images were not labeled, some of them responded preferentially to pictures of cats and some to pictures of people.

Le and his colleagues say that this system has discovered how to classify images containing cats and people, but what it appears actually to have learned are the statistical properties of the images. It learned to group pictures by similarity, because some of the hidden layers in the network were designed to treat similar input patterns in similar ways. Some of the output units corresponded to cats and people because many pictures of cats share some statistical properties. With 10 million training examples, there was ample opportunity for this correlation to be observed. Le and his colleagues identified the neuron that they supposed represented cats by presenting pictures and identifying the neuron that was most active in the presence of cats versus other things, but this is circular reasoning. The network did not know that this neuron represented cats. Le and his colleagues knew. As far as the network was concerned, it just responded to pictures and some of these pictures led to a certain pattern of activation on the outputs. The system did not create a category, it picked up a correlation. The designers, not the network, called that correlation a category. Even if the images were randomly distributed to output units, there would still be some that would respond more to cats and some that would respond more to people. It would not be correct to say that those units represented cats or that the computer learned its own categorization.

To be sure, Le's experiment is a massive undertaking, but it is easy to lose sight that with a billion variables, one can fit almost any function. It is also easy to oversell what this project accomplished. Autoencoding networks like theirs may help to improve the efficiency of using labeled examples, but they still cannot replace the labels entirely. Le and his colleagues applied the label after the network finished its learning, rather than during training, but it was they who applied the label.

Although deep learning networks are arguably more brain-like than multilayer perceptrons, they are still at risk of failing to live up to their hype. They are not a magic panacea. They surely solve some problems better than other networks, but they are not universal problem solutions any more than any other neural network or machine learning algorithm has been.

Deep neural networks excel at pattern recognition. Recognizing handwritten digits, or any kind of handwriting, has turned out to be a very difficult problem for computers to learn. Deep neural networks, though, have shown the most progress in this area. People write ambiguously and incompletely. Even people have a difficult time interpreting some handwriting, such as physicians' handwriting on prescription forms.

Deep neural networks have achieved human-level performance in things like recognizing traffic signs, segmenting the structure of neurons in electron microscope images, or identifying molecular structures that might lead to new drugs.

Neural networks more generally have been used in a wide variety of applications including:

- automobile guidance systems
- integrated circuit layout
- computer network anomaly detection
- helicopter transmission fault identification
- financial analysis
- stock trading strategies
- cancer cell identification and analysis
- facial recognition
- speech understanding
- evaluation of credit applications

Neural networks play an essential role in self-driving vehicles and in the system that learned to play go. They are critically important in the progress that computational intelligence has made over the last few years because they can do just the kind of pattern recognition that symbolic systems found so difficult.

One of the biggest limitations preventing neural networks from being more widely used is the large number of training examples that they need. Deep learning networks would not be possible in the absence of big data. The ability to provide high volumes of transactions, such as billions of Google searches, to these networks has made their style of learning possible. Training a deep learning network can take days or even weeks.

Training deep learning networks is also helped by the fact that they can be constructed from "modules." Different parts of the deep network can be trained independently of others. The unsupervised layers can be trained separately from the supervised layers, and the same learned structures can be reused with different supervised learning subnetworks without having to retrain the whole thing from scratch.

Still, training neural networks remains at least partly art as well as science. Complex networks simply have so many variables that they cannot all be set algorithmically. For example, there is no precise method to determine how many simulated neurons to put in a hidden layer.

Neural networks still need to be designed. At this point we have no algorithms and few heuristics to help us structure these complex networks. Deep learning networks can involve billions of parameters. The designer does not need to choose the value of each of them—that is what the learning algorithm does—but she or he does need to choose how those connections are organized. When neural networks come to be able to solve new kinds of problems, that solution comes first from novel ways of organizing the network. To this point, humans are still required to do that design. In this way, neural networks are not different from other forms of machine learning.

Adjusting the weights of even a deep neural network seems to me to be fundamentally different from the kind of changes that occur when we realize that we don't have to lay out the dominoes on the mutilated checkerboard to discover that it cannot be covered by a set of dominoes. Representing the checkerboard as a parity problem involving pairs of red and black squares is fundamentally different from representing it as a sequence of red and black squares. One is not a derivative of another. One is not a transform of the other.

The process that one engages in to make a decision is fundamentally different depending on how you represent the mutilated checkerboard. You do not have to lay out any dominoes to recognize that you cannot cover the board once you recognize that each domino has to cover exactly one red and one black square. The layout approach would not tell you how to solve a modified checkerboard that consisted of 126 red squares and 128 black squares, but the parity approach would tell you that if the number of red squares does not equal the number of black squares, then covering it with dominoes is impossible.

Not even deep neural networks are capable of learning representations that cannot be derived as transforms from the preceding layer or the inputs. Even deep neural networks implement searches through solution spaces, but the mutilated checkerboard problem requires a completely different solution space. We have yet to come up with a computational approach that can change solution spaces.

But maybe if we implemented an actual brain, we would be able to have a system that implemented humanlike intelligence. As John Searle (1990) said, brains cause minds. Maybe if we emulated a complete brain, we would automatically achieve general intelligence.

Whole Brain Hypothesis

If we had a running instance of Einstein's brain, would we have Einstein or even someone equally as intelligent? Taken to its extreme, there is a hypothesis that if only we had a complex enough neural network, we would have intelligence. If we had a neural network that mimicked the complete human brain, we would have a machine with the intelligence of a human, and maybe the personality of the person whose brain we are emulating. The neurons in this whole brain emulation would have to resemble more closely the actual operations of the mammalian neurons than artificial neural networks do today, but on the idea that two systems that can compute the same function(s) are equivalent, the argument goes that we would then have a mind.

Not every detail of the brain's neural processing can or needs to be emulated. In fact, there are some, such as Nick Bostrom, who argue that we really do not need to know much of anything about how the brain works; we just need to replicate its structure down to some level. The physiology of the brain cannot be ignored completely, but features and processes at the lowest level will have to be simulated rather than emulated. Sodium channels that allow ions to flow through the neuron's cell membrane are part of the mechanism by which neurons work, but computer chips do not have similar sodium channels, so this property, for example, will have to be simulated computationally. Advocates of whole brain emulation argue that some suitable level of abstraction can be found at which the computer can be said to emulate the brain rather than just simulate it.

Bostrom argues that molecular-level scanning of a brain will provide all of the necessary information we would need to replicate its structure and, therefore, its function, but even if we knew the complete structure of the brain, we would still need to have a solid understanding of its function to replicate it not as neurons but as circuits. I think that we need to know a whole lot more than that. For example, we have the full connectome (the full structure of each neuron and how it connects to each other neuron) of a small roundworm, *C. elegans*, but that information is not sufficient to explain even its behavior.

There is a lot of uncertainty packed into the whole brain emulation hypothesis. It assumes that we can get a suitable account of the brain dynamics at a suitable level of analysis. It assumes that we can understand the dynamics and structure of the brain sufficiently well that we can explain intelligence. It assumes that having that knowledge will allow us to implement intelligence. It assumes that we have the computational power to replicate those dynamics.

Of these assumptions, the easiest one to meet is probably the last one, that we can have sufficient computational power to emulate the brain. The human brain contains about 80 to 100 billion neurons with about 100 trillion synapses. Depending on the level of abstraction, we can reasonably describe the cycle time of the brain as being about 50 milliseconds. Relative to computers, things do not happen in the brain very quickly. For many purposes, if we can describe the state of the brain every 50 milliseconds, we are probably going to do a reasonable job of simulating it.

Emulation may require even higher temporal resolution. Neurons do not synchronize their activity to any kind of internal clock. They work asynchronously. They fire action potentials, for example, when the conditions are right, not when some overall clock says that it is OK. The asynchronous nature of neural activity makes modeling it with computers very difficult because computers tend to behave synchronously, according to a fixed clock—for example, one that ticks every 50 milliseconds. Still, a collection of fast enough computers might be able to simulate the asynchronous operation of the brain.

Because neurons operate more or less in parallel, the computing capacity of the brain has been estimated to be on the order 1 exaFLOP, which is roughly equivalent to a billion billion (10^{18}) calculations per second. It would probably be more proper to say that a computer simulating the brain in real time would have to achieve a speed on the order of 1 exaFLOP. The fastest known supercomputer in the world, Sunway TaihuLight, can achieve about 93×10^{15} (93 petaFLOPS), less than 1% of the estimated speed needed to do a brain simulation of the whole human brain. Both the US Department of Energy and the Chinese claim that they will have an exaFLOP computer prototyped by 2021. Even if it takes a few years longer than that, it would still be fair to say that the computational capacity to simulate (if not emulate) the human brain is reasonably likely in the next several years.

Simulations of some parts of the human brain have already been conducted. For example, Ananthanarayanan and colleagues (2009) used 147,456 processors and 144 TB of main memory to emulate a simplified version of a small part of the brain's visual cortex, consisting of 10^9 (1 billion) neurons and 10^{13} (10 trillion) synapses. This simulation took about 3 million core processing hours for 400 simulations. Their model mimicked the statistical properties of the neurons and connections in their selected brain area but did not attempt to replicate the complete set of neurons. One second of simulation, of even this small and simplified part of the brain, took about 200 seconds to run, using 7,500 hours of computer time distributed across all of these processors.

More recently, a team led by Markus Diesmann and Abigail Morrison simulated a network of 1.73 billion neurons and 10.4 trillion synapses. One second of neural activity took 40 minutes to complete using 82,944 processors and one petabyte of memory. Their model was not an attempt at brain emulation, but just a simulation of a small network of spiking neurons with what they called "biologically realistic connectivity." It did not replicate the organization of the brain it was simulating. It only preserved the statistical properties of the number of connections between neurons.

Despite considerable progress in mapping the fruit fly's connectome and elucidating the patterns of information flow in its brain, the complexity of the fly brain's structure and the still incomplete state of knowledge regarding its neural circuitry pose challenges that go beyond enough computational power to compute fly brain models.

Put simply, having computational power may be necessary to support brain emulation, but it is very far from sufficient. Compared to what we will probably need to know, neuroscience is practically at its infancy of scientific development. We know an awful lot about neuroscience today, but only a tiny fraction of what we will need to know in order to simulate, let alone to emulate, the human brain.

Part of the problem with the idea of full brain emulation is the assumption that the structure of the brain, how the neurons are connected to one another, is sufficient to replicate is functionality. Structure has something to do with the properties of how the neurons implement cognitive activity, but the structure is not at all sufficient. Earlier, I mentioned that neurons can change their roles over time, signaling one behavior at one time and another behavior at another time. If we don't know how individual neurons perform their tasks, what hope is there of modeling billions of them simultaneously?

The brain is not just a static structure, but a complex dynamic system that changes over time, as the individual matures and from second to second. We would, I think, have to map not just its structure, but its dynamic properties. Although we could, in theory, map out the structure of the brain down to the molecular level, we have no notion of how to map its dynamic properties, or even its current state. Assuming that a molecular scan would be sufficient to capture its state, presumably such a scan would not be instantaneous. By the time we mapped out part of the brain, the state of other parts are likely to have changed. The brain's state at the start of a scan would presumably be different from its state at the end. Dead brains do not change much over short periods of time, but it is doubtful that complete information about the brain's cognitive processes could be collected from a dead brain.

Mapping the state of a dynamic system such as the brain would face challenges analogous to those seen at the subatomic level in quantum mechanics. I don't think that the difficulty is caused by quantum mechanics, but the state of a neuron does, at least in part, depend on what statisticians call stochastic events. A stochastic event is one that has a certain probability of happening. At any point in time, it may or may not happen. We can predict stochastic events on average, but each individual event is difficult to predict with any accuracy. For example, the neurotransmitter molecules diffuse stochastically across the synaptic gap between the neurons so each molecule of neurotransmitter arrives at the receiving neuron with a certain probability after a random amount of time. Once at the receiving side of the synapse, the neurotransmitter molecules bind, again with a certain probability, with receptors.

We do not know how important it might be to preserve the details of the states of the neurons, the synapses, or the neurotransmitter molecules when doing the brain mapping. We really do not have a very good understanding of the roles each of these plays in implementing thinking in the human brain, so we are very far from being able to emulate, or even simulate, these features.

Experience plays an important role in determining some properties of the structure of the brain and, more importantly, the function of those structures. Hubel and Wiesel, working in the 1960s, studied the effects of sensory deprivation on the developing brain. For example, normal brains contain neurons that respond to one eye or the other. These cells are typically organized into "ocular dominance" columns where the neurons in alternating columns respond to (are dominated by) one eye or the other. A brain that develops with input from only one eye does not leave half of these columns unused; rather these neurons now both respond to the same eye.

In later experiments, neuroscientists found that if visual neurons were rerouted to the part of the brain that ordinarily processes auditory signals, they could still be used by the animal to navigate visually. Furthermore, a rewired auditory cortex—now responding to visual stimuli—shows the patterns of cellular response that are very similar to that typically seen in visual cortex. People who are blind from birth also show activity in the primary visual cortex when reading Braille.

But we do not have to go to prenatal development to find evidence that the brain can change in response to environment—inverting spectacles are enough. In 1896, George Stratton presented a paper describing an experiment that he did on himself in which he wore special goggles with prisms that inverted his visual field. After a short time, he was able to navigate around an indoor area while wearing these spectacles. Ivo Kohler, when a graduate student in the 1950s, found that within about two weeks of wearing inverted spectacles, the wearer had adapted to this large change in perception and could even ride a bike or catch a ball. This experiment indicates that even after maturation, the brain is still adaptable to fairly radical shifts in the inputs it receives.

Building exascale brain simulations may be useful as a research tool, but they would not automatically give us machine intelligence. There is no reason to think that if we built Einstein's brain, we would have the intelligence of Einstein or even a brain of similar intelligence. For example, identical twins are born with as close to identical brains as we can get. Identical twins reared together have similar IQ scores; their correlation is around 86%. But identical twins reared apart have a much lower correlation, around 76%. And if any two brains could be said to be reared apart, it would be a biological and a computational brain.

This is an imperfect argument; IQ tests are not perfect indicators of real intelligence, heritability is not a perfect indicator of how identical brains are, and so on. However, it does suggest that building a simulated brain may not be sufficient to produce high levels of intelligence. Presumably, the reason that IQs of identical twins reared together are more similar than those of twins reared apart is because twins reared together share more experiences. If experience is necessary to intelligence, then we have the issue of how do we provide this experience to computational brains.

One hypothesis for how to provide experience is to "upload" an actual person's mind from her or his biological brain into the computational brain. I have no more expectation that that would work than I have that the matter transporter from the old *Star Trek* series could actually be built, and for the same reason. The amount of data to be recorded and transmitted is just too huge. But in the case of capturing a person's mind, we also have sensor limitations. We cannot read the state directly of every neuron in the living brain, and we have dynamic limitations. A live brain is constantly changing. I would say that the likelihood of successfully uploading a mind is essentially 0.0.

In any case, there continue to be severe ongoing challenges to whole brain emulation. It is certainly not imminent. These challenges include:

- We don't have the computational resources to emulate a brain (this is arguably the easiest of the problems).
- We don't know enough about how the brain actually works to emulate it.

- We don't have methods to simultaneously record from the 100 billion neurons in an intact brain. We don't even know what we would need to record.
- We don't know what kinds of experience are essential for the brain to generate intelligence.
- We don't know enough about the effects of experience to successfully model it.
- We don't know enough about the dynamic properties of the brain.
- We know very little about how the brain stores memories.
- We don't know what consciousness actually means, how it is represented in the brain, or even whether it is important to intelligence.
- We have no idea how to record a person's personality or consciousness.
- We don't know how to describe and replicate the brain processes responsible for intelligence.
- We don't know how to apply machine learning to allow the brain to advance its functioning beyond relatively trivial capabilities.

There is some possibility that these barriers could eventually be overcome, but that is unlikely to happen in the foreseeable future. Computer resources will continue to improve, as will our knowledge of neuroscience. But the other barriers still seem pretty insurmountable.

Conclusion

Neural networks have taken computational intelligence a long way from the early expert system days and the idea that physical symbol systems are necessary and sufficient to produce intelligence. Artificial neural networks do not force the characteristics of the world to be cast into nice crisp categories. They do not require relationships among objects to be all-or-nothing. They allow continuous and gradual representations that appear to be more suited to actual situations. Because they involve so many parameters, however, they may be able to solve problems that can be cast as functions simply because of the large number of ways those parameters can be organized.

The notion of full brain emulation is attractive, but we are profoundly ignorant of many of the properties that would need to be modeled even if we had the computational capacity to achieve it.

Resources

Ananthanarayanan, R., Esser, S. K., Simon, H. D., & Modha, D. S. (2009). The cat is out of the bag: Cortical simulations with 109 neurons, 1013 synapses. In *Proceedings of the Conference on High Performance Computing Networking, Storage and Analysis*. New York, NY: ACM. doi:10.1145/1654059.1654124; https://people.eecs.berkeley.edu/~demmel/cs267_Spr10/Lectures/RajAnanthanarayanan_SC09-a63.pdf

Carpenter, G. A., & Grossberg, S. (2009). *Adaptive resonance theory* (CAS/CNS Technical Report No. 2009-008). Boston University. https://open.bu.edu/bitstream/handle/2144/1972/TR-09-008.pdf?sequence=1

Clark, W. A., & Farley, B. G. (1955). Generalization of pattern recognition in a self-organizing system. In *Proceedings of the March 1–3, 1955, Western Joint Computer Conference* (pp. 86–91). New York, NY: ACM. doi:10.1145/1455292.1455309

Driscoll, L. N., Pettit, N. L., Minderer, M., Chettih, S. N., & Harvey, C. D. (2017). Dynamic reorganization of neuronal activity patterns in parietal cortex. *Cell, 170,* 986–999.e16.

Givon, L. E., & Lazar, A. A. (2016). Neurokernel: An open source platform for emulating the fruit fly brain. *PLoS ONE, 11*(1), e0146581. doi:10.1371/journal.pone.0146581; https://www.ncbi.nlm.nih.gov/pmc/articles/PMC4709234

Grossberg, S. (1976a). Adaptive pattern classification and universal recoding: I. Parallel development and coding of neural feature detectors. *Biological Cybernetics, 23,* 121–134.

Grossberg, S. (1976b). Adaptive pattern classification and universal recoding: II. Feedback, expectation, olfaction, and illusions. *Biological Cybernetics, 23,* 187–202. In R. Rosen & F. Snell (Eds.), *Progress in theoretical biology* (Vol. 5, pp. 233–374). New York, NY: Academic Press.

Grossberg, S. (1988). Nonlinear neural networks: Principles, mechanisms, and architectures. *Neural Networks, 1,* 17–61. http://www.cns.bu.edu/Profiles/Grossberg/Gro1988NN.pdf

Harvard Medical School. (2017). Neurons involved in learning, memory preservation less stable, more flexible than once thought. https://www.sciencedaily.com/releases/2017/08/170817122146.htm

Hodgkin, A. L., & Huxley, A. F. (1952). A quantitative description of membrane current and its application to conduction and excitation in nerve. *The Journal of Physiology, 117,* 500–544. PMC 1392413 Freely accessible. PMID 12991237. doi:10.1113/jphysiol.1952.sp004764

Hopfield, J. J. (1982). Neural networks and physical systems with emergent collective computational abilities. *Proceedings of the National Academy of Sciences of the United States of America, 79,* 2554–2558.

Karn, U. (2016). An intuitive explanation of convolutional neural networks. https://ujjwalkarn.me/2016/08/11/intuitive-explanation-convnets/

Kohonen, T. (1984). *Self-organization and associative memory.* Berlin, Germany: Springer-Verlag.

Le, Q. V., Ranzato, M. A., Monga, R., Devin, M., Chen, K., Corrado, G. S., . . . Ng, A. Y. (2012). Building high-level features using large scale unsupervised learning. International Conference on Machine Learning. https://arxiv.org/pdf/1112.6209.pdf

Leonard-Barton, D., & Sviokla, J. (1988). Putting expert systems to work. Harvard Business Review. https://hbr.org/1988/03/putting-expert-systems-to-work

Minsky, M., & Papert, S. (1969). *Perceptrons. An introduction to computational geometry.* Cambridge, MA: MIT Press.

Minsky, M., & Papert, S. (1972). Artificial intelligence progress report (MIT Artificial Intelligence Memo No. 252). https://dspace.mit.edu/bitstream/handle/1721.1/6087/AIM-252.pdf?sequence=2

Morrison, A., Mehring, C., Geisel, T., Aertsen, A. D., & Diesmann, M. (2005). Advancing the boundaries of high-connectivity network simulation with distributed computing. *Neural Computing, 17,* 1776–1801. https://pdfs.semanticscholar.org/1bfb/a5de738f12afc279200e92f740f0d02cd964.pdf

Rochester, N., Holland, J. H., Haibt, L. H., & Duda, W. L. (1956). Test on a cell assembly theory of the action of the brain, using a large digital computer. *IRE Transactions on Information Theory,* 80–93.

Rosenblatt, F. (1958). The perceptron: A probabilistic model for information storage and organization in the brain. *Psychological Review, 65,* 386–408. doi:10.1037/h0042519

Rosenblatt, F. (1962). *Principles of neurodynamics: Perceptrons and the theory of brain mechanisms.* Washington, DC: Spartan.

Sardi, S., Vardi, R., Sheinin, A., Goldental, A., & Kanter, I. (2017). New types of experiments reveal that a neuron functions as multiple independent threshold units. *Scientific Reports, 7,* Article No. 18036. doi:10.1038/s41598-017-18363-1; https://www.nature.com/articles/s41598-017-18363-1

Searle, J. (1980). Minds, brains and programs. *Behavioral and Brain Sciences, 3,* 417–457.

Searle, J. (1990). Is the brain's mind a computer program? *Scientific American, 262,* 26–31.

Staughton, J. (2016). The human brain vs. supercomputers . . . which one wins? https://www.scienceabc.com/humans/the-human-brain-vs-supercomputers-which-one-wins.html

Werbos, P. (1974). *Beyond regression: New tools for prediction and analysis in the behavioral sciences* (Doctoral dissertation). Harvard University.

Widrow, B., & Hoff, M. E., Jr. (1960). Adaptive switching circuits. In *1960 IRE WESCON Convention Record* (Part 4, pp. 96–104). New York, NY: Institute of Radio Engineers.

6 Recent Advances in Artificial Intelligence

In this chapter we discuss some recent success stories in computational intelligence. IBM's Watson is important because it showed the public that computers could be capable of answering questions in a more or less naturalistic setting. Alexa, Siri, and other digital assistants are important because they extend the question answering to more practical, and more ubiquitous, applications. AlphaGo shows how formerly insurmountably complex problems could eventually be solved with innovative heuristics. Self-driving vehicles and poker playing show another kind of innovation, being able to deal with less structured problems and more uncertainty. All of these systems show important progress in terms of solving specific problems, but they do not get us significantly closer to achieving general intelligence.

Artificial intelligence has begun to disrupt broad swaths of industries that have previously required large teams of people. Computers and robots have started to displace even white collar workers in areas such as legal document review, medical diagnosis, and others. As disruptive as these changes may be, however, the hype surrounding computational intelligence is even worse.

Every day there are countless articles about how AI is transforming the world. Companies are jumping on the AI bandwagon; if they have any computational components to their products at all, they are advertising themselves as using artificial intelligence. Over a thousand companies claim to be AI providers. The flurry of hype is similar to that of the 1990s when seemingly every business transformed itself overnight to be a dot-com e-business.

Despite the hype, there is real value in computational intelligence that goes far beyond the capability to play complex board games. Cybersecurity, the process of protecting computers and computer networks from malicious intruders, for example, is an area in which machine learning has been very fruitfully applied.

The rate at which companies have been successfully attacked seems to grow every week. Cybersecurity is a war of attrition. As hackers get more sophisticated about how they hide their attacks, for instance, as attachments to emails, machine learning is getting more successful at identifying those attachments. The hackers use machine learning to disguise their malware, and security companies are using machine learning to identify it. Some recent security work, for example, has found malware enclosed in attachments, enclosed in attachments, and so forth, as much as 20 attachments deep. Each attachment layer obfuscates the contents of the layer below it by encoding the information, so it is very challenging to find the obscured malware without using machine learning.

Computational intelligence has long been active in financial settings. It has been used to uncover fraud and to identify potentially successful investments.

Health care is an active area receiving a lot of attention and investment for computational intelligence. Computational intelligence is used in a broad range of medical situations from dealing with issues of hair loss to cancer diagnosis.

Tech giants investing in health care include IBM, which is doing a lot of work to deploy its Watson technology in several health care areas; Philips, which is looking at health information from things like smart toothbrushes; and Google's parent company Alphabet, which is partnering with several universities to leverage deep learning for improving health care. Computational artificial intelligence in health care is growing at a high rate, measured in billions of dollars a year.

Electronic health care records provide an opportunity to do predictive modeling of future patient health based on the laboratory, diagnostic, and physician notes and other information in each record. For example, Riccardo Miotto, Li Li, Brian A. Kidd, and Joel Dudley used unsupervised learning to build a generalized representation of each patient that could be used to predict that person's future risk of one of 78 diseases. Their system was particularly accurate at predicting severe diabetes, schizophrenia, and some kinds of cancers.

Electronic health records suffer from the same kind of complications as text-based documents. The records contain many variables, which are often expressed in inconsistent forms. For example, type 2 diabetes can be identified by an A1C value greater than 6.5%, by a fasting plasma glucose level of 126 mg/dL, or by the presence of an ICD 9 diagnostic code of 250.00 or an ICD 10 code of E11.65, or the word "diabetes" can be mentioned in the record's clinical notes. All of this synonymy makes it difficult to associate any particular fFIeature in the record with any particular outcome.

In addition to predicting disease from the text of electronic health records, other projects have worked to diagnose cancer from images. Computers have been used to screen mammograms, for example, and to evaluate images of skin lesions for the presence of skin cancer. The accuracy of these systems has been found to be at least comparable to that of human radiologists and dermatologists.

The images to be interpreted are presented to the computer as an array of pixels—spots of light and dark and usually color. The images are different sizes, with variations in the position of the lesion, the lighting, and even the method with which the image was collected.

Andre Esteva and his colleagues trained a deep learning neural network (like those described in the preceding chapter) on a data set of 129,450 dermatologic images. They compared the performance of their system against the decisions of 21 dermatologists on diagnoses verified by biopsy for malignant carcinomas versus benign seborrheic keratoses; and for malignant melanomas versus benign nevi, a kind of birthmark or skin mole. Carcinomas are the most common forms of skin cancer, and melanomas are the most deadly. Each year in the United States, there are about 5.4 million new cases of skin cancer. Early detection of a melanoma means the person's chances of survival for five years can be as high as 97%, whereas at later stages, it is only about 14%. So, early detection of melanoma can be critical. Esteva and his colleagues found that their system was slightly better at distinguishing cancerous from benign lesions than the dermatologists.

A similar network was used to diagnose mammography images, also with high levels of accuracy. This system combined four images for each mammogram (one image from the top/bottom of each breast and one image from the side of each breast). Krzysztof J. Geras, Stacey Wolfson, S. Gene Kim, Linda Moy, and Kyunghyun Cho found that the higher the resolution of the images, and the larger the number of images used for training, the more accurate was the diagnosis.

Yun Liu and colleagues used a deep learning neural network to identify metastasis. Metastasis occurs when cancer spreads from one organ to another. There are many candidate tissues that could be the target of metastasis, each of which has to be examined carefully by a radiologist, which is labor intensive, expensive, and prone to error. Full-resolution microscope images of the tissue can be as large as 100,000 × 100,000 pixels. Their network recognized 92% of the tumors present in the images, which compares favorably with the 73% average of human pathologists.

This work on detecting cancer lesions, though promising, is still largely experimental. The models used involve many computational neurons, organized into networks with complex structures. It remains to be seen whether similar systems can be effectively deployed in more naturalistic situations with more variability in how the data are collected. Few radiology clinics are likely to have the kind of resources that have been deployed for some of these experiments. Despite some claims to the contrary, there is little danger that radiologists will soon be put out of business (see sources quoted by Siddhartha Mukherjee, 2017).

In the rest of this chapter I want to consider a few artificial intelligence projects that I think have had a major impact on their respective fields and on the development of computational intelligence. These projects are fundamentally academic exercises intended to expand the capabilities of what is possible in computational intelligence. They are more important for the ground that they have broken than for their direct commercial applicability.

The program Watson, which won so handily on the television game show *Jeopardy!*, has led IBM to produce a whole line of what they call cognitive computing. The goal is to use the kind of techniques that won on the game show to solve other kinds of problems. Perhaps more importantly, Watson caught the attention of the general public, who could see that computers were capable of doing tasks that were thought to require human intelligence. It became apparent that computers could behave in humanlike ways in real life, not just in science fiction.

Siri, Alexa, and similar programs take natural language understanding and question answering to new levels. Practically every smartphone has at least one virtual assistant program that can answer questions, make appointments, or do other simple tasks. These programs introduced into everyday life, perhaps in limited ways, the capabilities that Watson made so apparent. Now everyone could interact directly with the kind of capabilities that won on *Jeopardy!* and get answers from their cell phones or from similar so-called smart speakers on their end tables.

Google acquired the start-up DeepMind for $500 million in 2014. DeepMind was developing an artificial intelligence program to play the game Go, a game that was thought to be beyond the reach of artificial intelligence for the foreseeable future. Naturally, Google's interest ran deeper than just having a program that could play a game. Like IBM, they were looking for technology that they could generalize and apply in a broader context.

One of the most dramatic advances in artificial intelligence over the last few years was the emergence of self-driving cars. The Defense Advanced Research Projects Agency (DARPA) Grand Challenges had a profound impact on the development of artificial intelligence systems that could safely navigate vehicles. These vehicles, although not yet ubiquitous, have already changed how people drive. They portend to change the economics of trucking, warehouse logistics, taxi driving, and other areas. Their promise is to make driving simultaneously much safer and more efficient.

The last project that I want to highlight is an academic exercise in poker playing. It has not yet had a large public impact, but it is important for addressing a different kind of problem than that addressed by the other game-playing systems. Games like go, chess, checkers, and even *Jeopardy!* are all perfect-information games. All players have access to all information. Poker is different in that each player in the game has access to some private information—the cards in the player's own hand—that is instrumental to the outcome of the game. How an artificial intelligence system deals with this informational imbalance is a critical feature addressed by learning to play poker. Self-driving cars and poker-playing computers break significant new ground in computational intelligence.

Watson

In 2011, IBM's Watson competed on *Jeopardy!* and won against two human champion players, Brad Rutter and Ken Jennings. Here was a computer that could answer questions about real things in the real world. People were familiar with a computer storing vast amounts of knowledge, for example, with Google, but in their experience a computer returned a web page that might contain the information to answer their question. Watson actually returned answers.

Watson did not break any new theoretical ground, except in the scope of its accomplishment. It employed a wide array of state-of-the-art text processing tools. The range of questions that could be asked on *Jeopardy!* is enormous, and in order to win, Watson had to have a similarly wide-ranging knowledge, which it gleaned from over 200 million pages, more than 4 terabytes, of text and structured content, including Wikipedia. Watson also exploited several databases, dictionaries, taxonomies, and other reference materials.

For the show, Watson used a cluster of 90 servers, including 16 terabytes of memory. It could manage 2,880 processes simultaneously. It could process 500 gigabytes of text data per second, which is equivalent to reading a million books per second.

Watson employed more than 100 techniques to analyze the natural language, identify sources, formulate hypotheses, and score evidence and potential answers. One of these techniques, DeepQA, does not just look up its answers in a database of questions and answers; rather it analyzes the language of the question and the language of its sources to find potential matching answers. It then scores the answers using a variety of analytic techniques. It uses machine learning to learn how to weight these sources and analyses.

When presented with a question, Watson parses the question into keywords and sentence fragments, which it uses to look up related phrases. As a simple example, if the question asks who . . . , then the right answer must be the name, title, or description of a person. If the question asks when, then the answer must be a time. If the question includes pronouns, the referent of that pronoun has to be interpreted. It uses the result of this parsing process to look for matching information in its knowledge base.

The more of its analytic techniques that return the same answer, the higher is Watson's confidence in the answer. It then checks its answer against a database to determine whether the answer makes sense. Even with all of that, Watson still has a problem with very short questions that don't provide a lot of material for it to work with. Very famously, when asked to name the US city with two airports, one named after a World War II hero and one named after a World War II battle, Watson answered Toronto. The correct answer was Chicago, and, obviously, Toronto is not a US city. When Watson did make errors, its errors were not always of the sort that a human might make.

On the other hand, people do make errors that are somewhat similar to answering Toronto. Answer quickly: How many animals of each kind did Moses take on the ark? Most people immediately answer "2," but the correct answer is actually 0. It was Noah, not Moses, who took animals on the ark (according to the Bible).

DeepQA looks at question answering as a process of generating hypotheses from its analysis of the question and its available knowledge. It then ranks these hypotheses in light of the evidence.

In this, Watson is very much in the spirit of physical symbol systems or expert systems. It differs from these in its use of machine learning, in its comparison of competing hypotheses (as opposed to following a logical path through a series of choices), in its use of processes to disambiguate questions, and in its ability to extract information from unstructured sources.

Traditional expert systems rely on hand-coded rules that reason either from evidence to conclusion (given this evidence, what can you conclude?) or from conclusion to evidence (what evidence do you need to find for this conclusion to be true?). DeepQA's natural language processing and machine learning automates the process of matching knowledge approximately against the questions it receives.

IBM's work on Watson was undertaken for a number of reasons beyond the marketing glow of being able to claim a big win in a television game show. They wanted to create a general-purpose natural language processing and knowledge representation and reasoning system that could be reused in many different domains (such as medicine). They wanted a system that could gain its knowledge both from structured sources, such as databases, and from unstructured sources, such as text. They wanted a system that would learn quickly and answer quickly while also being highly accurate.

They succeeded brilliantly in creating an effective question answerer. If winning at *Jeopardy!* were the criterion for the Turing test, then they would have passed beyond any reasonable doubt. Although the topics about which it could answer questions were very broad, it was still a specific system, solving a specific problem. It could not go on to play go or chess based on the training it received and the processing power it exploited. Even broad knowledge of facts was not enough to make it generally intelligent. It could not reason beyond the scope of the processes that were specifically created to win at *Jeopardy!*.

Watson helped make it apparent that there was great potential in computational intelligence, but it did not really get us much closer to artificial general intelligence. Ken Jennings, one of the two human *Jeopardy!* contestants competing against Watson, was premature in welcoming his new computer overlords. If they ever show up, Watson will not be among them.

Although IBM is hopeful that Watson will help lead it to exploit the future of computational intelligence, to this point, IBM is apparently still facing the challenge of making it a profitable business (Strickland, 2018).

Siri and Her Relatives

Digital voice assistants, such as Siri, Alexa, and Google Assistant, are applications, often hosted in phones or smart speakers, that allow users use their voice to interact with various services. Two things are remarkable about these applications: first, they are able to understand voice commands, rather than forcing users to type in what they want, and second, they are able to perform sometimes sophisticated tasks in response to the requests that they receive. Rather than simply looking up some result in a database table, these systems can perform actions. They play music, find a time when two people can meet, or recommend a restaurant and reserve a table. They keep track of the context of a request, such as your food style preferences, and use that information. Most of this interaction is still relatively basic, but they are constantly "learning" new "skills." I put learning in quotes here because it is not clear that they actually use machine learning to acquire these skills.

Spoken language provides a "natural" way for people to interact with various kinds of technology. People are used to having conversations with others, and it is a small step to have a conversation with a digital assistant. The assistants provide a uniform method of interacting with a broad range of specialty services without having to learn the peculiarities of each separate system. They incorporate machine learning to discover their users' preferences, neural networks to interpret speech, and potentially other kinds of artificial intelligence to organize and execute their actions.

Although most of these virtual assistants are still fairly primitive, their conversational interface encourages users to personify them and often to attribute more intelligence to them than would otherwise be merited. On the other hand, their ability to perform tasks, even if the tasks are fairly simple ones, means that users can be freed of the burden of these tasks.

Some of the tasks performed by virtual voice assistants are controlled by following a programmed set of rules. The rules consist of conditions and actions. For example, if the stock MSFT drops 10 points, then sell it. Such a rule requires the system to identify which stock is meant, track its value, and then take an action when the conditions are met. It needs a memory to know what the price was before and compare it with the current price. It needs a way to execute a stock trade. Another rule might be to copy a picture to a Google Photos account when you "like" the picture on Facebook, while also automatically tagging the photo for the people in it. Or, it could add milk to your shopping list when the milk carton is empty.

Virtual assistants can tell you what the weather will be in your location (which they recognize automatically) and remind you to take an umbrella when that is appropriate. They can order a ride from ride-sharing services. They can provide medical advice from WebMD or other sources.

These systems cannot do everything, of course. Their ability to answer general information questions, for example, is still limited. They still lack common sense.

In many ways the more significant development in digital virtual assistants is their ability to recognize speech. Although speech recognition systems are now common and readily available, how that capability was developed is interesting as an example of the use of computational intelligence and machine learning.

In 1952, researchers at Bell Labs developed a system, called Audrey, that could recognize isolated digits spoken by a single speaker. The first commercial application to do speech recognition was the IBM Shoebox, introduced in 1962. It could recognize 16 words: the digits zero through nine, "plus," "minus," "subtotal," "total," "false," and "off." Notice that although the word "false" is included on this list, its opposite, "true," is not. Their method could not distinguish, I infer, "true" from "two."

Starting in 1971, DARPA (at the time it was called ARPA), the name did not yet include "Defense") funded a multiyear research effort to develop speech recognition that could handle a thousand-word vocabulary. One of the products of that effort was Carnegie Mellon University's HARPY system, which could recognize 1,011 words with reasonable accuracy.

Continuous speech recognition, where you do not have to pause between words, came to be available in 1990, with Dragon System's DragonDictate. After training for an extended time on a single voice, the $9,000 Dragon program could transcribe speech with fair accuracy. In 1997, they released Dragon NaturallySpeaking, which then cost only $695. After 45 minutes of training the system to recognize the speaker's voice patterns, it could manage a single user's continuous speech at about 100 words per minute.

In 2008, Google launched voice search on the iPhone. Soon after, they enabled other programs to use Google's speech-to-text conversion. Now there are many systems, some of them freely available, that are quite accurate at speech recognition. In 2011, Apple released the first version of Siri on the iPhone. The previous year, Apple had acquired the core of Siri in an acquisition of Siri Inc., a spin-off of SRI International, a research institute that does much of its work for the US Defense Department.

From the 1950s to today, speech recognition has gone from being able to recognize 16 isolated words spoken by a single speaker to being able to recognize millions of words in 110 languages. Speech recognition is a difficult problem, and the progress over the past 60 years has been powered largely by two kinds of developments: improved representations of the speech problem and improved availability of speech and text examples. In this, speech recognition is a model for much of the improvement in computational intelligence—better representations and more data.

Speech recognition is so familiar that viewing it as a computational challenge may seem odd. But in reality, speech signals are extremely ambiguous. Using speech recognition to drive an intelligent agent is even more difficult. To build a voice-controlled agent requires that we go from the vibrations in air to correctly completing an action.

This process requires several steps (each of which is ambiguous):

- Map acoustic (sound) events to phonemes.
- Map phonemes to words.
- Map words to intents.
- Map the intent to action.

The relationship between speech sounds and the language they represent is itself very complex. Speech sounds are the acoustic patterns, the physical vibrations of the air. The acoustic properties of a speech sound can be represented by its time-varying power spectrum (see chapter 5). A power spectrum represents the amount of energy contained in an acoustic signal at each of several frequency bands. A time-varying spectrum represents the amount of energy in each frequency band as the pattern changes over time. In a computer, the time varying spectrum is estimated using the fast Fourier transform, described earlier. In the ear, the equivalent transformation is produced by the physical characteristics of the cochlea.

Those time varying spectra have to be translated into the linguistic representation of speech sounds, called "phonemes." English contains about 42 phonemes (depending on dialect). These range from long a sounds, /A/, such as in "hay," to /ks/ sounds, such as the x in "axe" or the /z/ sound as in "nose."

The linguistic representation is needed because there is not a simple or direct mapping between acoustic signals and their corresponding phonemes. For example, the /p/ in the syllable, "pi" (sounds like pea) is actually the same acoustic pattern as the /k/ sound in the syllable "ka" (Cooper, Delattre, Liberman, Borst, & Gerstman, 1952), but people hear them as completely different.

The same acoustic event can correspond to more than one linguistic event. Phonemes are linguistic categories. When learning to speak a language, a child must learn to associate the acoustic patterns she hears with the phonemes that are appropriate for that child's language. That association depends on context, as shown by the pi versus ka experiment. In general, the sound that comes before and the sound that comes after a specific sound can affect how that sound is interpreted into phonemes. Speech recognition systems must manage this ambiguity.

The HARPY system, mentioned earlier, used a graph search algorithm to identify phonemes. It kept track of the alternative phonemes or words that were consistent with recently received sound patterns and then chose the sequence of phonemes that was most consistent with the sequence of sound patterns. HARPY represented the initial parts of the speech recognition system as a network of constraints and then navigated this network to identify the best guess as to what the sound pattern represented.

Once the phonemes in a speech sound have been identified, the next step is to map those phonemes to potential words. That mapping is also ambiguous. For example, a certain set of sounds could be interpreted as "visualize whirled peas" or as "visualize world peace." Many words are pronounced identically in ordinary use, such as "ladder" and "latter." Homophones are distinct words that are pronounced the same, such as "wear" and "where." So there is substantial ambiguity between the phonemes and the spelling of words.

Strong progress in speech recognition was made when a language model was added to the acoustic model mapping sound patterns to phonemes. Language models represent the probability of a word in the context of the words that have preceded it. The word "ladder" is much more probable than "latter," for example, in sentences that contain the phrase "climbed the corporate . . . ." The opposite is true, following the phrase, "given a choice, he chose the . . . ." Language modeling uses machine learning to estimate these probabilities and then uses those probabilities to interpret the words that were actually said.

The creation of these statistical models was made easier by the fact that very large amounts of text came to be stored in a way that could be easily accessed by computers. The Dragon models, for example, were trained for a specialized subject matter by presenting large amounts of text data. The text did not have to be spoken because its role was just to signal which words were more or less likely in the context of preceding words. Google, of course, had nearly unlimited amounts of text in practically every written language and billions of queries along with their results. When they introduced Google Voice, furthermore, they gained access to large amounts of speech examples with widely varying accents and styles as well.

Further ambiguity comes from the use of pronouns, such as "he" or "it." In a sentence like "Find me an online store that has a pashmina shawl and buy it," the system has to determine that "it" refers to the shawl and not to the store. This sentence also illustrates another problem that the natural language understanding computer must solve, that of speaker intent.

In a sentence like, "He poured the milk from the bottle into the bucket until it was empty," the "it" must refer to the bottle. But in the corresponding sentence "He poured the milk from the bottle into the bucket until it was full," the "it" must refer to the bucket. This kind of ambiguity cannot be resolved by the structure of the sentence. The two sentences have exactly the same structure. It can only be resolved by real-world knowledge that pouring changes the contents of two containers and the one being poured from becomes more empty while the one being poured into becomes more full.

Even more difficult are sentences like "The police would not stop drinking" or "They are cooking apples." In the first sentence, who is drinking? In the second, are they apples that are good for cooking, or is someone cooking the apples?

In recent years, speech understanding has been facilitated by the use of deep learning neural networks. Starting around 2012, Google began using Long Short-Term Memory Recurrent Neural Networks to do the Google Voice transcriptions. They trained these deep learning networks using discriminative training, requiring the system to contrast sounds, not just learn each phoneme independently. This discriminative training takes advantage of the fact that the pattern of successive phonemes depends on the preceding phonemes that have been identified. The successive phonemes have to match real words, and the sequence of words has to match patterns actually seen in the language. The sequences of both phonemes and words has to make sense, in other words. Using the voice message transcriptions offered as part of Google Voice, Google could get examples of speech sounds and text and get some feedback when the system got it wrong and the user suggested alternative transcriptions. Being voice mail, these sequences were about as natural and conversational as could be had.

Even after the words have been properly recognized, the voice agent must still determine what it should do with the message. Transcribing voice mail stops once the text has been written down. On the other hand, virtual agents are expected to do something with what they have understood. The original intent of the project that eventually led to Siri was to create a personal assistant that would, for example, organize email, calendars, documents, and schedules; perform some tasks; and facilitate other communications.

Among the tasks that an agent might perform, one could be to make a travel reservation. A sentence like "I need to book a flight to New York on July 7" might seem fairly unambiguous, but even this simple sentence presents significant challenges. The phrase, "I need" requires interpretation that the speaker's intent is to actually travel to New York. It requires knowledge that travel requires a ticket and a reservation on an airline. The computer needs to know where the speaker is currently. It may need to know other things about airline preferences and so on (such as preferred flight times). Does "book" in this context mean to accuse someone of a crime, or does it mean to reserve a seat? It may need to know when the speaker wants to return. If the agent is rather limited, then some of this ambiguity disappears in that the agent is designed to interpret the ambiguous phrases in only a limited range of ways.

But, just as speech recognition was enhanced by developing representations that took more context into account, intent recognition is likely also to be enhanced by taking more context into account. The information needed to be successful may not be fully contained within the request itself but may depend on outside sources of information that also need to be integrated.

For a typical one of these personal assistants, a question, a command, or a query begins with the user's voice request. A compressed version of the voice recording is sent to the system's server (most of the work occurs on the service provider's server rather than on the phone). Automatic speech recognition translates the voice recording into text. The type of query is then identified (action request, command, search query). If it is a command to the phone, the appropriate command may be sent then or, if it involves Internet or other knowledge resources, such as databases or other users' calendars, those resources may be accessed. An answer is selected from the available responses and sent back to the user.

AlphaGo

The game of go has been mentioned several times already. Go is a strategy game for two players, played on a board consisting of a 19 × 19 grid (361 positions, compared to checkers or chess with 64 positions), on which black and white pieces called "stones," are placed. One player places the black stones, and the other places the white ones. Each player tries to surround more territory on the board than his or her opponent.

The players take turns placing the stones, one per move, on the "points" (intersections) of the grid. Once a stone is placed on the board, it cannot be moved but can be captured by surrounding it with opposing stones. Captured stones are removed from play. The game has no set ending. It can be ended by resignation or by the players deciding not to make any more moves. The player with control of the larger territory wins.

The complexity of playing go is not due to the complexity of the rules—there are only a few—but is due to the number of possible moves that need to be considered at each move.

Every position on the grid can be in one of three states. It can be occupied by a black stone, occupied by a white stone, or empty. There are 3^{361} possible positions, of which about 1.2% are legal. Therefore, there are about 2.08×10^{170} legal positions in go. The actual number is 208,168,199,381, 979,984,699,478,633,344,862,770,286,522,453,884,530,548,425,639,456, 820,927,419,612,738,015,378,525,648,451,698,519,643,907,259,916,015, 628,128,546,089,888,314,427,129,715,319,317,557,736,620,397,247,064, 840,935.

This huge number of combinations was thought to put the game out of reach of conventional algorithms. There are just too many possibilities to consider. By comparison, chess is estimated to have about 10^{123} possibilities, which is a miniscule fraction (a decimal point followed by 47 zeros before a 1) of the possibilities for go. Some people are fond of saying that there are more positions to either of these games than there are atoms in the visible universe (about 10^{80}).

Like other artificial intelligence approaches to games, playing go can be described as the process of navigating through a space starting from an empty board. A brute-force algorithm would assess each potential move and choose the one that had the highest expected value (ultimately, the highest probability of leading to a win), given the current configuration of pieces on the board. But the huge number of possible moves at any point in time and the complexity of computing each move's expected value makes this approach infeasible. The breakthrough came from some clever design of heuristic methods that could be effective at selecting a subset of potential moves so that they would not all have to be evaluated.

Chess and go are considered perfect-information situations because there is no uncertainty about the state of the game at any point in time. Each player may be uncertain about what the other player will do on future moves, but each one knows perfectly what the players have done to that point. Both players know the state of the game, the rules, the locations of all of the pieces, and so on. If a player selects a move, there is no uncertainty about how that choice will affect the state of the game.

The complexity of the search process is determined by its breadth and its depth. Breadth is the number of legal moves per turn, and depth is the number of subsequent choices in the game following that move. In chess breadth is about 35 (there are about 35 legal moves at any point in time) and depth is about 80 (each side makes about 80 moves in a game). A complete analysis of a chess position, then, would evaluate 35 potential moves for how each of those moves would change the state of the game over the next 80 similar choices. In go, breadth is about 250 and depth is about 150, too many to consider all of them. Heuristics are needed to play chess or go.

Heuristics can be used to select the stronger moves for evaluation, putting aside other moves. If we cannot evaluate all of the moves, then a better player will be the one that evaluates moves that are more likely to be effective and does not waste time considering moves that are unlikely to lead to a winning outcome.

Heuristics are acceptable when playing a game like go or chess because human players are not able to fully evaluate each move either. Rather, chess players who have been studied tend to rely on patterns of chess pieces that they have seen before and go players claim to use aesthetic judgment to decide which moves to make. The quality of play depends on the quality of the heuristic selection process.

AlphaGo is a go-playing program that beat one of the world's best go players, Lee Sedol, four games out of five in March of 2016. AlphaGo ran on 1,920 standard processors and 280 graphical processing units, distributed over a number of data centers. Lee used only his brain. The graphical processing units in this case were not used to manipulate graphics but to do the complex matrix operations needed for the heuristics.

Players of chess and go learn the patterns of pieces that have previously led to success. AlphaGo also learned to play go, in part, by studying past games. Given enough pictures of a cat, a computer can learn to identify cats. Given enough examples of go games, a computer should be able to learn to play go.

AlphaGo was trained on billions of go moves. It used deep neural networks to learn how the game is played. Some of these example games were played against human players, but many of them were played against other go-playing computers, including other versions of AlphaGo.

As AlphaGo played itself, it kept track of which moves were more successful at controlling territory on the board. It played millions of games against itself, gradually improving, and abstracting properties of these patterns in the same way that similar neural networks can abstract visual properties from millions of images. Recall, from chapter 4, that Arthur Samuel used a similar strategy to help his system learn to play checkers.

Although trained on millions of go games, AlphaGo was not limited to merely mimicking those games. It did not just memorize the previous games; it abstracted principles. These principles did not necessarily correspond to those a go expert would describe, but its substantial experience allowed it to identify these principles nonetheless. Principles, in this context, are statistical regularities.

In game 2 of the match against Lee Sedol, the computer made a move that no human player would be likely to make. In fact, AlphaGo estimated the probability of that move at 1 in 10,000. Once AlphaGo's stone was placed, Lee could see quickly that this was an unexpected move that he did not recognize. But Lee was also learning, apparently, because in game 4, he made an unexpected move himself, from which AlphaGo never recovered.

AlphaGo represents some innovative machine learning techniques that may have application in other computational intelligence situations. One of the most interesting is having the system learn from playing itself. Using deep neural networks to abstract patterns is also a critical insight with broader applicability. Its search algorithm and the methods by which it chose its policies may also be of interest.

AlphaGo is viewed by some as an artificial intelligence program that learned to improve itself. A self-improving AI scares some people. But I think that this fear is completely misplaced. All machine learning programs are self-improving, and AlphaGo is not different in this regard. It does what it was designed to do.

It has no capability of transferring the knowledge of game play it gained from its games of go to any other game, let alone any other kind of task. AlphaGo and its ability to beat Lee Sedol may be an important milestone in the development of artificial intelligence, but it is not a departure in kind from the machine learning that came before it. It learned how to search a problem space and find novel paths within that space, using heuristics it was provided by its human designers.

Self-Driving Cars

Another dramatic artificial intelligence project in recent years is the emergence of self-driving vehicles. According to *Wired* magazine, over

263 companies were working on self-driving vehicle technology in 2018. According to ABC News, there were 52 companies approved to test self-driving vehicles in California alone in 2018.

Part of what spurred the interest in self-driving cars was the DARPA Grand Challenge, offering a million dollars to the first team that could field an autonomous vehicle capable of traveling over an unrehearsed off-road course. DARPA is the research agency tasked with finding innovative solutions for the problems faced by the US military.

The first Grand Challenge competition took place in March of 2004. Teams were informed only hours before the start of the event what specific course they were expected to follow on a 142-mile trail through the Mojave Desert. Of the 15 teams that started the race, none of them succeeded in driving more than 7.5 miles of the route.

Carnegie Mellon University's Humvee drove too close to a cliff edge, where it spun its wheels until a tire caught fire and the vehicle was shut off. Another vehicle started the race, but its Global Positioning System (GPS) malfunctioned, and it drove around in circles. A vehicle fielded by a team from Palos Verde High School crashed into a concrete barrier near the start of the race. The competition that first year did not exactly turn in spectacular results, but even 7.5 miles was a major accomplishment.

DARPA repeated the challenge again in 2005, increasing the prize to $2 million, and the results were dramatically different. This time 23 teams entered a 132-mile race and five of them finished. The winning vehicle, named Stanley, was built by the Stanford team.

In principle, the strategy for a self-driving vehicle sounds simple. It must know where it is, know where it is trying to get to, avoid obstacles, obey traffic rules, and choose the best course of action.

Anyone who has used Google maps or one of the other navigation programs knows that there have been big improvements in map quality and in route planning since these applications first became available. In the 2005 Grand Challenge, however, the vehicles had to do their own route planning because they were navigating through the desert and they were seldom on paved roads.

The vehicles used GPS for identifying where they were when it was available. GPS is a system of satellites. Under the right conditions, a GPS system can identify its location to within a few feet. Given the rocky, hilly, desert environment, though, GPS signals were often lost.

Stanley, the winning vehicle, used several different kinds of sensors, including GPS, lasers, video, and radar to identify its position and direction. It used lasers and radar to identify obstacles that it would have to avoid. Accelerometers, gyroscopes, and wheel sensors were also used to identify the position and pose (for example, whether it was tilted at a dangerous angle). With this suite of sensors, Stanley could determine its position to within a couple of inches.

To avoid obstacles, the vehicle must detect them at a range that allows it to take evasive action or stop before colliding with it. At short range, up to about 22 meters (about 24 yards), Stanley used lasers to detect obstacles. The lasers were useful, therefore, at speeds up to about 25 mph. For longer range obstacle detection, in order to travel at higher speeds, Stanley used radar and stereovision. All of these data were processed by seven shock-mounted laptop computers in the vehicle's trunk.

Unlike many of its competitors, Stanley was not programmed with a set of rules. Instead, it was given the opportunity to learn how to drive during the months leading up to the race. Part of the machine learning process for Stanley was to have a human driver control the car, navigating only through drivable terrain. The data from the paths that the driver actually took could then be labeled as drivable and other areas as not drivable. This approach means that some of the terrain to the left and right of the vehicle's path is mislabeled as not drivable when, in fact, it may be flat usable terrain. It does ensure, however, that drivable terrain is correctly labeled and can easily be used. It also meant that the team had a ready source of data for training their machine learning algorithms using supervised learning.

Contributing to the difficulty of classifying surfaces into drivable and nondrivable, the appearance of the road is affected by factors that change over time, such as the material (for example, asphalt or concrete, the darkness of the asphalt), the lighting (for example, the angle of the sun, the degree of cloud cover), camera wobble, and dust, both in the air and on the camera lens. Even a flock of birds suddenly taking off in front of the vehicle could change the appearance of the road. As a result, the road-following module had to be adaptive to a range of highly changeable conditions.

One of the insights that led to Stanley's success was the recognition that in the real world, sensor data are always contaminated by "noise." The sensors would get shaken around. Dust would interfere. Rocks and tunnels would obscure GPS signals. Fortunately, the factors that affect visual interpretation of camera images are different from the factors that affect radar and laser range finding. So some of the ambiguity of the visual interpretation could be reduced by the other sensors. I believe that this redundancy and sensitivity to different kinds of noise was instrumental in the success of the Stanley system.

Another key insight was the recognition that Stanley had long-range sensors (radar and video) and short-range sensors (lasers) and that when it drove forward, objects detected by the long-range sensors could eventually come into the range of the short-range sensors. Stanley could learn by tracking the predictions made by the long-range sensors and then using the short-range sensors to teach the system about the usefulness and interpretation of the signals obtained from the long-range sensors.

There has been a lot of progress in self-driving cars since the DARPA Grand Challenge. One motivation is safety. Every year in the United States, human-driven cars end up causing about 30,000 deaths, or one death for every 90 million miles driven. Self-driving cars have the opportunity to reduce this number of fatalities and injuries substantially. There have been two known deaths from an autonomous vehicle, a Tesla that ran into the side of a truck while the car was on "autopilot" and a pedestrian killed while crossing the street in front of an Uber autonomous test vehicle.

Reportedly, drivers have been using Tesla's autopilot for about 300 million miles. We do not have enough data to determine whether 1 in 300 million is a reasonable estimate of the likelihood of a fatal accident from an autonomous vehicle, but it seems promising. We have less information about the number of miles driven by Uber's self-driving cars (*The New Yorker* magazine estimated it at 3 million miles in 2018; Sheelah Kolhatkar, 2018), and their program may not be as advanced as Tesla's or Waymo's. From these few deaths, it is difficult to extrapolate how autonomous vehicles will do as their use becomes more widespread.

The success of Stanley in the 2005 Grand Challenge, as well as the subsequent success of self-driving vehicles, derives in large part from the way that Stanley represented the problem to be solved. Stanley represented the problem as a machine learning one. Representing the relationships among long-range and short-range sensors was another critical representational decision as was training the car by driving on the kind of terrain that it would encounter during the Grand Challenge.

Poker

The game of poker presents computational challenges that are absent from many other games. In poker, the opposing player has information, the cards she has been dealt, that the computer does not have, and that person could lie (bluff). The appropriate strategy for a computer playing poker depends on its estimate of what cards the other player or players might have. These cards were randomly dealt, but there is also information that can be gleaned from what the player has done up to that point in the game.

One of these programs, DeepStack, plays a variation of Texas hold'em called "heads-up no-limit Texas hold'em." It is a two-player game with computational complexity comparable to that in go. Each hand progresses through four rounds, during which cards are dealt and the players can bet. During the "preflop" round the players are each dealt two cards face down, the private cards. Each player knows only the cards he was dealt. At this point, all the player knows is that the other player was not dealt these specific cards.

During the "flop" round, three additional cards are dealt face up. These cards are publicly known and can make up part of either (or both) players' hands. The next round is called the "turn," when one more card is publicly dealt face up. The fourth round is called the "river" when one additional face-up card is dealt. Each player can make a poker hand out of a subset of the five public cards dealt face up and the two private cards.

The face-up cards provide perfect information to the two players, but they are also identical for both players. But the preflop, face-down cards are known directly only to the player to whom they were dealt. Each player knows his or her own hole cards (those dealt face down during the preflop round) but is uncertain about the cards held by the other player.

The players bet before the first round and at the end of each round A player can raise (increase the bet), fold (surrender the pot to the other player and end the hand), or call (meet the bet proffered by the other player).

The opponents' betting behavior is publicly available to both players and can be used to estimate the strength of the opponent's hand. But either player could bluff. They can bet as if they have a very strong hand, counting on their opponent to fold, despite the fact that the opponent may actually have the better hand. Because of bluffing, public betting behavior is only imperfectly related to the strength of the player's hand. There is no limit to the amount of money that a player can bet in no-limit Texas hold'em, except that a player cannot bet more money than he has.

Both players try to infer the strength of their opponent's hands from the imperfect cues available in the opponent's betting behavior. Each player's betting behavior, in turn, depends on that player's estimates of his or her own chances as well as the chances of his or her opponent. These estimates can vary widely from one round to the next as more cards are revealed and as more betting occurs.

Imperfect-information games are important because they are so much more complex than perfect-information games. An effective player has to choose a strategy for each hand that will be effective in light of the uncertainty of what cards the opponent has. The opponent may have a 9 of hearts and a 3 of spades as his or her hole cards. Or the opponent may have the 9 of hearts and the 9 of spades. The computer may be able to estimate the probabilities of all of the different combinations but then must prepare for these differing probabilities.

An opponent's betting patterns reveal information about what the opponent is holding and about what the opponent thinks the other player is holding. The reasoning is recursive, wherein each player affects the decisions made by the other, which changes the other's betting behavior, which changes the first player's betting, and so on. In poker, the rules are not much more complicated than in go. Each hand involves only a few moves, but the number of potential states and the uncertainty about those states makes the game extremely challenging.

The full complexity of the game makes a complete analysis impractical. Rather, DeepStack breaks the game down into components. Instead of estimating the complete game from each point (each bet), it computes a fast approximate estimate of the value of each move. It forms this estimate by having the computer play during training from random poker situations, and selects them by probabilities of a good outcome.

Computer programs are superior to human players in estimating the probabilities of each card and of the strength of achievable hands given the face-up cards and the cards in the computer's own hand. The opponent cannot have either of the cards held in the computer's private hand. The cards that are face up, on the other hand, are in both players' hands. If the king of hearts card is showing face up and the computer has the king of spades, for example, then the opponent cannot do better than to have two kings in his or her concealed cards. The computer cannot do worse than have a pair of kings.

The poker-playing method used by DeepStack contains three main parts: (1) a local strategy estimate for the current public state (what is known about the visible cards and the betting behavior so far), (2) the set of actions that it contemplates, and (3) a look-ahead method to estimate the likely consequences of each of these potential actions.

Before each bet, the computer recalculates its strategy in light of the current state. The computer searches its solution space, as it would for a complete information game, but because of the uncertainty concerning its opponent's hand, it must consider a wider range of possible states than just the one that will yield the highest return. DeepStack uses two deep neural networks, which are trained to estimate the future states of the game without having to exhaustively recalculate them before each bet. One network was employed during the flop round (when three face-up cards are dealt) and a different one during the turn round (when an additional face-up card is dealt). Each neural network contained seven hidden layers of 500 neurons each.

The turn neural network was trained by playing 10 million randomly generated turn states. That is, for each randomly generated game, the play was "simulated" up until that point. The network evaluated the various potential actions for that specific configuration of cards shown and bets and then played the game through to the end, with a restricted set of potential actions: fold, call, bet the current pot, or bet all-in (all the player's current money). The inputs to the network were the pot size, the publicly visible cards, and a categorized estimate of the cards that the opponent might have in the hole. The flop network was trained with an additional 1 million randomly generated flop states, using the estimates from the turn network to value each potential action.

After training, DeepStack's performance was measured by playing 44,852 games against human players. The high variability in this kind of poker game requires some specialized statistics, but the results were favorable for the computer. On one measure, which is the value won per game relative to the size of the minimum bet, DeepStack won 0.492, where a professional poker player would consider 0.05 to be a sizeable margin. A break-even player would achieve 0 on this measure, so the computer did pretty well. In other words, if the minimum bet were a dollar, DeepStack would win an average 49.2 cents per game, against a field of good, but perhaps not the best, players.

Conclusion

The success of these computational intelligence programs over the last several years is due largely to three factors: first, the availability of good-quality training data, which made large-scale supervised learning possible; second, the use of pattern recognition to characterize the problem solution space; and third, clever ways to represent the problem being solved.

The operations of the artificial intelligence system were represented as approaches to recognizable categories of patterns as opposed to rules governing responses to specific named inputs. The systems were designed to largely discover their own patterns from a candidate representational space designed by their creators.

Among the most prominent machine learning approaches to pattern recognition and classification are the so-called deep neural networks. These systems consist of several layers of simulated neurons that allow the input data as originally represented to be transformed into patterns that may be more computationally tractable. They can generalize from the patterns provided during training to related patterns that they had not seen. In essence, they abstract the input patterns that they receive into derivative patterns.

Another factor that contributed to the success of some of these important AI programs is the use of one machine learning system to train another one. The self-driving cars, for example, used the laser range finders to help train the visual analysis system. AlphaGo used one version of itself to play against another version of itself to provide training examples. DeepStack used a random system to create poker hands that could be learned by its deep neural networks. This technique addresses a major bottleneck in some of the most successful forms of machine learning, the need to use large numbers of labeled training examples. Certain kinds of learning problems lend themselves to this kind of adversarial training. When combined with reinforcement learning, they can be very powerful tools in the training of effective machine learning systems. But this process comes with a risk.

Garbage in, garbage out. If one system is working against another system, there is no guarantee that it will be successful when working against a different—for example, more natural—adversary. One system may merely learn the flaws inherent in the other, and those flaws may be different or absent from other systems or players.

Finally, these examples are important because they show how the kinds of problems that are addressed by machine learning can be extended. There are many advantages to studying problems that are easy to understand (even if they are complex to solve), but the world does not consist of only these well-structured problems. Playing go is important because it is an example of a situation where simple representations would leave a machine lost in thought. Advanced, creative representations, including insightful heuristics, made an intractable problem tractable.

Self-driving cars and poker playing take machine learning into realms of imperfect information. They can be addressed because new representations have been invented that allow for uncertainty. Speech-based assistants combine uncertain, ambiguous inputs with uncertain intentions. The range of actions and intentions that may ultimately be addressed by these systems is potentially enormous. The representations that allow these systems to be useful are still evolving.

All of these examples and many others, though, also share an important property. These programs' success depends on some designer finding an appropriate and useful representation of the problems that they face. This representation has to transform the problem from one that may be unsolvable into one that can be resolved within the capacity of modern computers. Their intelligence comes largely from the cleverness of the designer. Achieving artificial general intelligence will require finding a way to replicate that representational creativity, which, so far, has relied on human capacities and talents.

Resources

Cooper, F. S., Delattre, P. C., Liberman, A. M., Borst, J. M., & Gerstman, L. J. (1952). Some experiments on the perception of synthetic speech sounds. *The Journal of the Acoustical Society of America, 24,* 597–606. doi:10.1121/1.1906940; http://www.haskins.yale.edu/Reprints/HL0008.pdf

Esfandiari, A., Kalantari, K. R., & Babaei A. (2012). Hair loss diagnosis using artificial neural networks. *IJCSI International Journal of Computer Science Issues, 9,* 174–180. https://pdfs.semanticscholar.org/6217/c168b99db35605144169d3efc20ab195a2cf.pdf

Esteva, A., Kuprel, B., Novoa, R. A., Ko, J., Swetter, S. M., Blau H. M., & Thrun, S. (2017). Dermatologist-level classification of skin cancer with deep neural networks. *Nature, 542,* 115–118. doi:10.1038/nature21056; https://www.nature.com/articles/nature21056.epdf

Geras, K., Wolfson, S., Kim, S. G., Moy, L., & Cho, K. (2017). High-resolution breast cancer screening with multi-view deep convolutional neural networks. https://arxiv.org/abs/1703.07047

Kolhatkar, S. (2018, April 9). At Uber, a new C.E.O. shifts gears. *The New Yorker.* https://www.newyorker.com/magazine/2018/04/09/at-uber-a-new-ceo-shifts-gears

Kolochenko, I. (2017). How artificial intelligence fits into cybersecurity. CSO. https://www.csoonline.com/article/3211594/machine-learning/how-artificial-intelligence-fits-into-cybersecurity.html

Kubota, T. (2017). Deep learning algorithm does as well as dermatologists in identifying skin cancer. Stanford News. https://news.stanford.edu/2017/01/25/artificial-intelligence-used-identify-skin-cancer/

Lee, J. (2013). OK Google: The end of search as we know it. Search Engine Watch. https://searchenginewatch.com/sew/news/2268726/ok-google-the-end-of-search-as-we-know-it

Liu, Y., Gadepalli, K., Norouzi, M., Dahl, G. E., Kohlberger, T., Boyko, A., . . . Stumpe, M. C. (2017). Detecting cancer metastases on gigapixel pathology images. https://arxiv.org/pdf/1703.02442.pdf

Metz, C. (2016). In two moves, AlphaGo and Lee Sedol redefined the future. *Wired.* https://www.wired.com/2016/03/two-moves-alphago-lee-sedol-redefined-future/

Miller, A. (2018, March 21). Some of the companies that are working on driverless car technology. ABC News. https://abcnews.go.com/US/companies-working-driverless-car-technology/story?id=53872985

Miotto, R., Li, L., Kidd, B. A., & Dudley, J. (2016). Deep patient: An unsupervised representation to predict the future of patients from the electronic health records. *Nature Scientific Reports, 6,* Article No. 26094. https://www.nature.com/articles/srep26094

Moravčík, M., Schmid, M., Burch, N., Lisý, V., Morrill, D., Bard, N., . . . Bowling, M. (2017). DeepStack: Expert-level artificial intelligence in heads-up no-limit poker. *Science, 356,* 508–513. doi:10.1126/science.aam6960; https://arxiv.org/pdf/1701.01724.pdf

Mukherjee, S. (2017). What happens when diagnosis is automated? *The New Yorker.* https://www.newyorker.com/magazine/2017/04/03/ai-versus-md

Naimat, A. (2016). The new artificial intelligence market. https://www.oreilly.com/ideas/the-new-artificial-intelligence-market

Silver, D., Huang, A., Maddison, C. J., Guez, A., Sifre, L., van den Driessche, G., . . . Hassabis, D. (2016). Mastering the game of go with deep neural networks and tree search. *Nature, 529,* 484–489. doi:10.1038/nature16961; http://airesearch.com/wp-content/uploads/2016/01/deepmind-mastering-go.pdf

Stanford Encyclopedia of Philosophy. (2014). Speech perception: Empirical and theoretical considerations. https://plato.stanford.edu/entries/perception-auditory/supplement.html

Strickland, E. (2018). Layoffs at Watson Health reveal IBM's problem with AI. IEEE Spectrum. https://spectrum.ieee.org/the-human-os/robotics/artificial-intelligence/layoffs-at-watson-health-reveal-ibms-problem-with-ai

Tromp, J. (n.d.). The number of legal go positions. http://tromp.github.io/go/legal.html

7 Building Blocks of Intelligence

Up to this point we have been concerned with forming a preliminary definition of general intelligence and then taking stock of the ways that it has been approached in human cognition and computational intelligence. This chapter begins a discussion of the kinds of resources that might be available to overcome the current limitations. A main focus of this chapter is on the idea that cognition does not flow in only one direction. What we perceive and what we think is affected by context and by expectations. The chapter continues with a discussion of how language is both a problem for intelligence and a contributor to intelligence, and it concludes with a discussion of common sense.

Long before ancestral brains were doing anything that we would currently recognize as intellectual achievement, they were evolving the capability to sense, perceive, and act in their environment.

Figure 6 shows a moth resting on the bark of a birch tree. Blue jays have little trouble locating the moths, but people often have to search a long time to find it, if they ever do. How dumb are we? We do not typically think of tasks like finding moths as part of intelligence, but if our diet depended on finding these moths, we might have a different idea.

This task is conceptually not much different from projects designed to distinguish photographs containing cats from those that do not. Here, we want to distinguish trees that hold moths from trees that do not, and we want to locate the moth on that tree. It may take a computer millions of examples before it can reliably find the cats, but blue jays would probably starve if they required even thousands of training examples.

Even well-experienced blue jays find it easier to detect cryptic (camouflaged) moths immediately after they have found a similar one. This

Figure 6
A Catocala moth on a birch tree. https://www.researchgate.net/publication/282230039_Selective_Attention_Priming_and_Foraging_Behavior/figures?lo=1. Used by permission. See figure 9 if you cannot find the moth in this picture.

tendency, called a "search image," was first proposed by Niko Tinbergen in 1960. Finding one example of a particular prey species makes it easier for the predator to find another example of that species than it is to find an example of another species. Both prey species may be equally hard to find, and both may be equally common in the predator's diet, but still, finding one makes finding more of the same easier. There is some kind of short-term attentional effect that helps the blue jay or other predator to find more examples of the same thing. Experiments have supported Tinbergen's idea and the notion that there are contextual, attentional factors that affect perception. The idea of a search image is one example of how perception is more than picking up stimulation from the environment but is an active process governed by attention and expectations.

Perception and Pattern Recognition

Humans and many other animals have evolved specialized neurons for sensing the environment. The most familiar of these are probably the retinal receptors in the eye and the cochlear hair cells in the inner ear.

Light reflected from objects in the environment is projected through the lens of the eye landing upside down on the mosaic of photosensitive cells, the retina, at the back of the eye. Originally it was thought that the retinal cells merely transmitted a signal corresponding to the amount of light that struck them. What neuroscientists found, however, was that pattern processing begins in the retina. Ganglion cells in the retina combine inputs from many photoreceptors and transmit patterns to the rest of the brain.

The pattern processing in the retina is the first step in a complex cascade of feature detectors that continues through several brain layers. Around 1959 Hubel and Wiesel began reporting on cells in the visual cortex that would respond to bars of light in specific parts of the visual field. They and subsequent neuroscientists followed these patterns of selective response to more complex processing in other areas of the brain following the primary visual cortex.

More recent research has also found what looks like top-down (brain to retina) changes to the responses in the early layers of visual processing, for example, depending on attention. Visual processing does not proceed in only one direction, but the action of lower layers is affected by the action of higher layers further along the chain of visual processing. These top-down processes focus attention on some things at the expense of others, including increased activation of some neurons and suppressed activation of others.

As discussed in previous chapters, hearing sounds starts with the mechanical action of the eardrum and the bones of the middle ear. The bones of the middle ear transmit the sounds to the cochlea, which provides a mechanical frequency filter bank. Specific parts of the basilar membrane in the cochlea respond to specific frequencies.

As in the eye, the ear also shows evidence of both top-down (brain to ear) and bottom-up (ear to brain) processes. The ear includes inner hair cells, which sense the frequency pattern of sounds, and outer hair cells, which provide mechanical feedback and amplify some frequencies at the expense of others.

Other sensory systems seem to perform in similar ways, involving both bottom-up and top-down activity. Sensors receive signals from the environment, transduce those signals into neural activity, and represent those signals in a spatially distributed manner. But perception is then actively modified by events happening later in the processing chain—there is feedback. Perception is more interactive and more object oriented than previously thought, further eroding the distinction between symbolic (object) and so-called subsymbolic (sensory) processes. This kind of active feedback process is potentially critical to the functional intelligence of organisms including humans.

Gestalt Properties

The perceptual system seems to have evolved for dealing with objects rather than for dealing with specific sensory patterns. The Gestalt psychologists, primarily Kurt Koffka, Max Wertheimer, and Wolfgang Köhler, in the 1910s and 1920s identified a set of principles that seem to be important in determining just what people perceive as objects: proximity, similarity, continuity, closure, and connectedness. These properties demonstrate again that perception is not a simple product of the stimulation that strikes the sensory surface, but it is a constructive process that uses other sources of information to recognize the objects that could produce such a sensory experience (see figure 7).

The neurobiology of brains has long been recognized as important to understanding intelligence, both machine and biological. Since Donald Hebb (1949) first proposed his learning rule, there has been significant cross-fertilization of artificial intelligence research and neuroscience. Hebb's learning rule can be summarized as saying "Neurons that fire together, wire together." More formally, "When an axon of cell A is near enough to excite a cell B and repeatedly or persistently takes part in firing it, some growth process or metabolic change takes place in one or both cells such that A's efficiency, as one of the cells firing B, is increased." This is one of the first neuropsychological explanations for how learning occurs in the brain.

Hebb's rule and its descendants continue to be among the most influential principles in neural networks. The feature detection and feature processing of deep neural networks is inspired by what we know of the feature processing of the visual system. But the relationship between brains and networks is still more metaphorical than literal. We don't think that we are mapping the structure of the visual cortex when we build a deep neural network.

Ambiguity

Ambiguity further challenges our view of perception. Analogous to the ambiguity we discussed in the preceding chapter in the context of speech

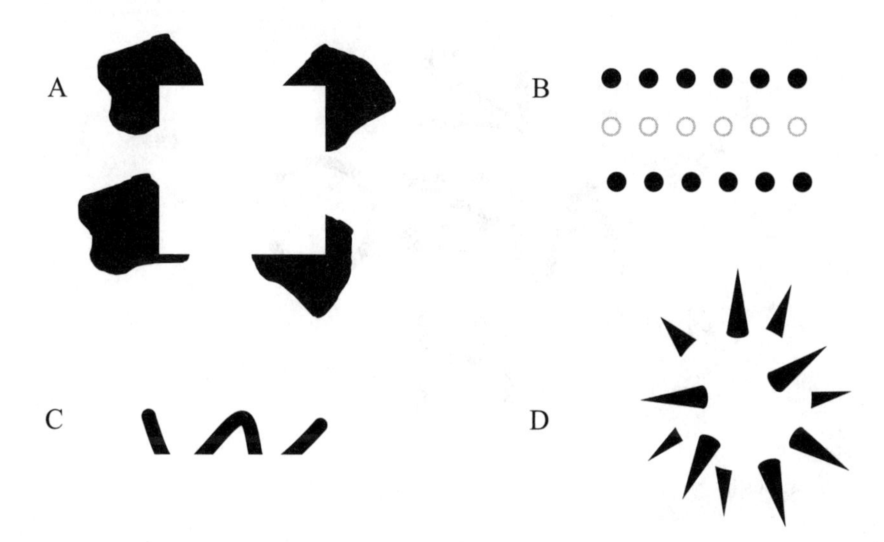

Figure 7
Gestalt principles. Four drawings that represent some of the characteristics of Gestalt perception. A shows an illusory figure. There is no square actually present in A. B shows the principle of similarity. Items that are physically similar tend to be seen as part of the same object. C shows good continuation. The three segments are usually seen as part of the same object, such as a partially submerged sea serpent. D is usually perceived as a ball with spikes, but there is no ball.

recognition, there is no simple mapping between visual scenes and the objects that are recognized in those scenes. The same visual pattern in the center of the display shown in figure 8 can be identified as either a B or as the number 13, depending on whether you look at it starting from left to right or starting from top to bottom.

The ambiguity also extends to the sounds that we hear and to the words that we use every day (see chapter 6). Although early views of artificial intelligence focused on using symbols that were analogous to words, it turns out that words themselves are not constant symbols of anything. Many people are familiar with the ambiguity of words like "bark" or "bank" or "strike," but the widespread ambiguity of other common words and their degree of ambiguity may be more surprising.

As an exercise, I looked up each word from the sentence in table 4 in a dictionary. The numbers below each word indicate the number of definitions I found for that word in the dictionary. If you combine each of these

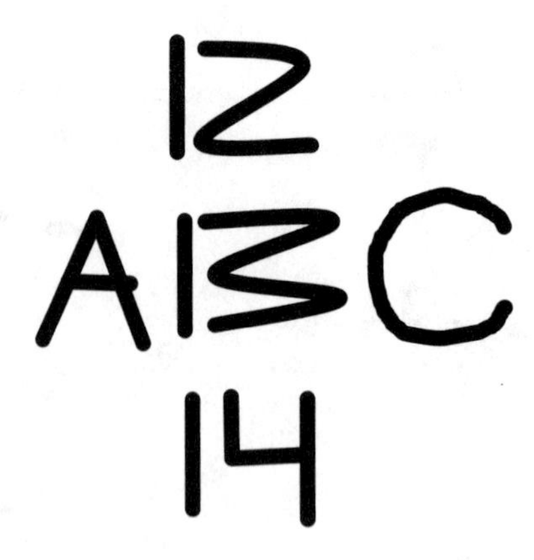

Figure 8
Ambiguous B versus 13.

definitions, you get almost 8 quadrillion (7,788,584,618,680,320) possible interpretations of this sentence, yet few people even notice any ambiguity.

Table 4

The	com-panies	have	agreed	to	a	brief	delay	in	imple-menting	their	agree-ment.
37	14	39	17	54	62	20	8	84	8	7	9

To be sure, the number of dictionary definitions is an imperfect measure of linguistic ambiguity. But it does suggest the qualitative level of ambiguity of ordinary language.

We seldom notice the ambiguity, even in simpler sentences like "She ate her lunch next to the bank," because the words do not contribute independently to the meaning of the sentence. Consider "She ate her lunch on the bank," versus "She ate her lunch in front of the bank." She could have been on top of a building (or a piggy) when eating her lunch according to the first sentence, but we are much more likely to interpret it as meaning she was at the side of a river. In the second sentence she could have eaten her lunch in front of a river's edge, but we tend to interpret it as meaning in front of a financial institution.

Any system that depended on the atomic nature of words would have problems. The notion of atomic words is the idea that the word "cat" has the same meaning in any sentence that contains that word. "The cat wore a hat," "The cat sat on the mat," or "The cat smiled at Alice" all employ the same symbol, and the meaning of that symbol is intended to be consistent in the three sentences. The meaning of the sentence is thought to be a composite of the meaning of the words in it. Sentences involving words like "bank" or "bark" were thought to be rare exceptions to the idea of atomicity and compositionality. Although these words looked like the same word, they were really, so the thinking went, different symbols that just happened to be indistinguishable. In contrast to this idea, it turns out that ambiguous words are more the norm than the exception.

The notion of using context to help a machine to understand the meaning of words is a key idea in the open-source project Word2Vec and a number of other projects that seek to represent the meanings of words by how those words are used in the language. Each word in the vocabulary is represented in these systems by its co-occurrence pattern in the text. With what words does this word occur? Words that are similar in meaning, it turns out, tend to occur in similar contexts, that is, with certain other words. The word "lawyer," for instance, is likely to appear with many of the same texts that the word "attorney" appears with. These two words have similar meaning, and by embedding their representation in similar contexts, the computer can abstract some of that meaning, so that when a user searches, for example, for the word "lawyer," she may also get documents that do not mention "lawyer" but do mention "attorney." More generally, the ambiguity of words reflects again the influence of top-down processes on the way we perceive and respond to events in the world.

Intelligence and Language

Despite the ambiguity, one of the most important inventions that support human intelligence is language. Language is just as much an instrument of human intellectual accomplishment as are maps, computers, mathematics, and fire. In all cases, these instruments of intelligence function to facilitate the mental or computational operations to support human accomplishment.

Before the widespread use of calculators, mathematicians used slide rules. A slide rule is a combination of two narrow boards, one of which could slide relative to the other. It could be used, for example, to multiply or divide two numbers just by sliding one of these boards relative to the other. The slide rule made it possible to easily perform complex mathematical operations, like multiplication, division, and exponentiation, that would have been difficult or time-consuming to perform mentally or on paper. Slide rules made certain mathematical operations practical, making the people who used them smarter than before. The slide rule is an instrument of artificial intelligence.

These instruments of intelligence, including language, make it easier to generate information, store it, transform, retrieve it, or otherwise use it. They impact our intellectual capabilities in a way similar to how software packages allow simple desktop computers to perform powerful and useful tasks.

The instrumental value of language has long been recognized by developmental psychologists. Language is not identical with thought, but it helps to structure and organize it in ways that are essential for the kind of intellectual capabilities we usually associate with intelligence.

Among the psychologists recognizing the importance of language in the development of intellectual capabilities were Jean Piaget and Lev Vygotsky. According to Piaget, the intellectual development of the child occurs in four regular stages.

During the sensorimotor stage (infancy to age two years), children's knowledge of the world is derived from the actions that they perform in it and their sensory experience of the world. Basic language is acquired toward the end of the sensorimotor stage.

The preoperational stage (two to seven years of age) follows. The child's knowledge is controlled primarily by the external world. The child has limited ability to focus on more than one aspect of an object or problem at a time. Thinking at this stage is prelogical, verging on magical. The child has a difficult time understanding that other people have a different point of view.

The concrete operational stage (seven to eleven years of age) begins to allow the child to reason logically and systematically. The focus is still on concrete objects. The child begins to be able to engage in reversible thinking and begins to recognize that each person has a unique view of the world. This logical and systematic reasoning is typical of the kind of skill we often call "intelligent."

The formal operational stage follows (adolescence and beyond). During this stage the person is capable of demonstrating logical thought and abstract concepts. Problems can be assessed in a logical and systematic matter. Not everyone seems to reach the formal operational stage, however.

The sensorimotor stage extends from birth to the onset of language. Early in this period, children are limited to basic reflexive actions, but by the end, they show the beginnings of symbolic thought. They quickly progress from babbling to being able to express themselves, using one or two words at a time.

Children do not have the ability to understand logic at the start of the preoperational stage. They have only very limited ability to mentally manipulate information. They begin to pretend at this stage, however, showing some evidence of symbolic thought. For example, they begin to play social games with roles, have pretend tea parties, and play house. During the second half of this stage, children's linguistic constructions become more sophisticated as does their ability to reason.

Children in the concrete operational stage are able to incorporate inductive logic. They can infer general principles from specific examples but may have difficulty with deductive logic. They may have difficulty using a general principle to predict the outcome of a relevant event. They are mostly limited to reasoning about concrete objects, actions, and situations.

Eventually, many children reach the formal operational stage, where thought is freed of the concrete constraints that characterized earlier stages. They can effectively use language related to abstract concepts and reason about hypothetical situations. In the formal operational stage, children achieve the highest levels of intellectual capabilities that we associate with intelligence.

Piaget did not particularly pursue the specific role of language in these developmental stages. Another developmental psychologist, Lev Vygotsky, on the other hand, was more interested in the role that language plays in intellectual development.

Vygotsky noted that a child faced with a difficult problem may get verbal coaching from a nearby adult to better accomplish the task. Later, when the adult is no longer around, the child may use similar speech, either aloud or internally, to replay those instructions and accomplish the task without the adult. People often ask themselves, "What would my father do?" in difficult situations. People talk to themselves about what they are doing when they are working on hard problems. The speech seems to help them to structure the action.

According to Vygotsky, language helps children to augment their prelinguistic cognitive capacities, such as attention and associative learning, with new abilities for focused attention and symbolic thought. They use the language to structure their thoughts. Cultural systems, such as language, instruction, science, and books, increase a person's intelligence. During early stages of development, this instruction is also a social activity in which children acquire their culture through interaction with others, including through explicit adult instruction. The process of development is to transition from behavior that is regulated by others to behavior that is regulated by the self. In Vygotsky's view, this transition is largely through internalization of formerly external instruction. Early on, thought is nonverbal and language is nonintellectual. Over time, thought incorporates more of the properties of verbal activity and speech becomes more rational.

I believe that it is fair to think of the early stages of development, as characterized by Piaget and Vygotsky, as providing the foundation for what Kahneman called "System 1 thinking." Kahneman did not characterize his distinction in terms of child development, but the characteristics of his System 1 are not very different from those attributed to young children by Vygotsky and Piaget. Both Vygotsky and Piaget focused on intellectual intelligence as the end point of cognitive development, but there is no reason to think that the nonlogical and impressionistic capabilities had to disappear as the more logical and intellectual capacities emerged.

Another developmental view of the emergence of intelligence is the argument that intelligence emerges from an exceptional ability to learn by analogy; the possession of symbol systems, such as language or mathematics; and the relationship between the two by which the ability to exploit analogy is amplified by the ability to use language. This is the position advocated by Dedre Gentner and others.

According to Gentner, among the capabilities that distinguish human intelligence relative to other species' are:

- the ability to draw abstractions from particulars
- the ability to maintain hierarchies of abstraction
- the ability to concatenate assertions and arrive at new conclusions
- the ability to compare and contrast two representations and how they are different
- the ability to invent and learn terms for abstractions as well as particular entities

In particular, she sees a progression during child development from first responding to physical similarity, followed by the ability to judge similarity based on selected features, to eventually the ability to focus on relational or conceptual similarity. In Gentner's view, children move from perceptual similarity to conceptual similarity. This is another example of top-down influence in thinking. We not only derive concepts from specific examples (induction) but those concepts then affect how we perceive and judge events.

For example, in one experiment, Gentner showed children pictures of a knife and a watermelon. The kids were told that the knife was the "blick" for the watermelon. "Blick" is a made-up nonsense word. She then showed the child an axe and a tree and said that the axe is a blick for the tree. Finally, she showed the child a piece of paper, scissors, a pencil, and another piece of paper and asked which of these three objects was a blick for the paper. The scissors represented relational similarity. Like a knife, it cuts. The pencil represented thematic similarity. Pencils are used to write on paper. The second piece of paper represented nominal similarity. It was the same as the paper. She found that four- and six-year-old children chose the scissors as the blick for the paper, but younger children simply chose randomly. Until they were four years old, children had a difficult time understanding the functional relationship among a knife, an axe, and scissors.

The ability to engage in these relational analogies is enhanced by the ability to talk about the relationship. Naming a relational pattern increases the probability of seeing it in other situations that are perceptually distinct from the original. Relational language creates symbolic pairings that might not otherwise occur, Gentner says. Relational terms can also help to focus on properties that are specific to the point of view expressed by the relation. When asked to name the properties of a pet, for example, people will mention different things than when asked to name the properties of a carnivore or a good mouser.

Offering people a discount for early registration results in fewer takers than offering them a late payment penalty for late registration, even if the dates and the price differences are identical, as Kahneman found and as we discussed earlier. Highlighting a particular relation focuses the person on some properties (winning) or others (losing) and can result in different choices for formally identical situations.

Language allows each new generation to learn from past generations, even though the meaning of terms can drift over time. Language augments the ability to hold and manipulate concepts and sets of concepts.

Words help to structure how we think about things. The names we apply to things influence how we think about them. A strong version of this idea is the Sapir-Whorf hypothesis named after Edward Sapir and Benjamin Whorf. The idea is that the concepts we have of the world are determined by the categories codified in our native language. In its extreme, it says that we have a very difficult or perhaps even impossible time thinking about concepts that we cannot express directly in our native language.

The bulk of evidence collected in the process of investigating the strong Sapir-Whorf hypothesis has been negative. As a strong notion, that we can only think of things for which we have words, it is very clearly false. Lawrence Barsalou, for example, found that people are quite capable of making up ad hoc categories on the spot (name things to take with you if your house is on fire). We do not have a word for such a concept, but Barsalou found that such ad hoc categories have the same kinds of cognitive properties that more traditional categories have. For example, people can select prototypical members of that category.

Still, it does seem clear that the words we use for things influence how we think about them. In the aftermath of the hijacking on September 11, 2001, the airlines banned anyone from taking knives on a plane. They replaced all of the metal knives in the public spaces of airports and on the planes with plastic versions. They did not, however, ban forks. I argue that the knife ban was because knives were categorized as weapons, but forks were not, even though a fork or even a spoon can be just as deadly a weapon as a knife.

George Lakoff explores this idea of linguistic categorization and its impact on thought in his book *Women, Fire, and Dangerous Things*. Lakoff's point, along with Barsalou's, is that there are categories that cannot be described by the similarity of the features of the objects that make up that category. What is the common feature that makes up the category of things to sell at a garage sale?

The failure of feature similarity to define categories is not limited to ad hoc ones. Ludwig Wittgenstein first talked about this in the context of what makes up a game. He argued that there was, at best, family resemblance among games. What is the similarity among baseball, tic-tac-toe, charades, and solitaire? Yet, we categorize them all as games.

Even family resemblance may give too much credit to physical similarity as the basis for categorization. According to Medin (1989), the similarity between two items, on the standard view, should depend on the number of features that they share in common versus the number that they share or don't share. But, in fact, any two items share an infinite number of features. A chicken dinner and a boy scout will both fit in a Volkswagen, they both take up space, neither is named Sam, they both weigh less than 300 pounds, they both weigh less than 301 pounds, . . .

The problem does not go away if we assume that categories are represented in the mind by examples of that category. In computers, this would be called a "nearest neighbor classifier." Unfortunately, nearest neighbor classifiers still depend on similarity to determine the nearest neighbor.

Tversky and others have shown that the features that would be compared in a similarity-based categorization are too flexible to be the basis of conceptualization or category membership. Even if we limited the comparison to features that were salient, that people paid attention to, and that they mentioned, similarity would still be a weak means of identifying items with their categories. The features that are mentioned by people depend strongly on the context in which the similarity judgment is being made. Tea may be considered a typical drink when talking about secretaries having lunch, but not when talking about American truck drivers taking a break.

Just as two items may share an infinite number of features (foxes and squirrels both have hearts and skin, but these features are rarely mentioned), a given object may be in a potentially infinite number of categories. Rolf may be a dog, he may be a male, he may live in New Jersey, he may be a living thing, and so forth. He may be one of the things that Sophie loves.

The features that one selects, or weights more heavily in judging similarity, are affected by the objects being compared but are not determined by it. Similarity is not sufficient for human categorization; rather, human categorization seems also to be affected in a top-down manner and then to affect the dimensions by which we judge similarity. Categories affect the features by which we compare at the same time that features affect the categories to which we assign objects.

The recognition that similarity is not sufficient to manage categorization raises problematic issues for machine learning or computational intelligence. It implies that computers must have knowledge that is not given by stimulus features directly. They must have contextual and categorical knowledge.

In existing computational intelligence methods, the feature selection comes from the representations chosen by the designers of the computer system. If pixels are used to represent images, then similarity is determined by the overlapping pixels in two images and the classes are determined by the similarity of the pixels. Some systems transform the information in the pixels mathematically to extract higher-order representations that are more abstract than the raw pixel images. These transformations are determined by the structure of the pixel data and of the layers in the neural network.

Computational intelligence systems are not at this point capable of deciding for themselves what their representations should be. As a result, the features appropriate to one problem may not be useful when applied to other problems. These are implicit decisions made when structuring a machine learning problem, but at the same time, we also do not have a good account of how humans select the features that they consider when categorizing in relatively unconstrained situations. Better understanding of how humans judge similarity would likely be extremely helpful to building out more powerful computational intelligence systems.

Still, it is interesting to note that the way that psychologists have characterized human cognitive development as a process of layering more abstract and rational processes on top of basic perceptual processes has a ready analogy in the layers of deep neural networks. It is also interesting to note that human judgments about similarity depend strongly on the context in which the similarity judgment is made. Context is emerging as an important part of some kinds of machine learning, such as in Word2Vec. These studies also imply that labels, in this case words, can play an important role in categorization that is more subtle than simply providing the right label for a category. Finally, the emphasis on analogy and prototypes may suggest further useful developments in machine learning. We will come back to these ideas in chapter 12.

Common Sense

Common sense is what we call everyday reasoning about nonacademic subjects. If you hear that John has a job, you infer that he works most days, that he earns money, and that he probably has a boss of some kind. Common sense represents facts about an individual's world and the relations among those facts that are not contained directly in the representation of the objects being reasoned about.

Common sense is essential to many kinds of natural language understanding situations. For example, if I say, "I took the tube to Marble Arch," you could understand me to say that I carried a cylindrical container to a place or thing called Marble Arch, but you would be much more likely to understand me to say that I took the London subway train to the station called Marble Arch. How we interpret a sentence can depend on information that is not contained in the sentence, or even in the text surrounding the sentence. Interpretation may depend on real-world knowledge. Recall the sentence from chapter 6 about pouring water from a bottle into a bucket until it was empty. You need common sense to decide whether the word "it" refers to the bucket or the bottle.

It is difficult to say what facts and relations constitute common sense. But presumably, common sense is what allows us to function in a world where everything is not explicitly specified. Common sense can be thought of as the set of facts, prejudices, background assumptions, and convictions that are implicit in our everyday reasoning about people, their intentions, and their actions. Knowing that John fell down after drinking for several hours, we can infer that he was drunk. Knowing that Nicole is Martha's aunt lets us infer that Martha is Nicole's niece, and that Nicole's husband (if she has one) is Martha's uncle. A sibling of Nicole or a sibling of Nicole's spouse is one of Martha's parents.

Common sense lets us reason about cause and effect, motion, personal relations, force, and energy and quantities, among others. It helps people to describe, predict, assess, and explain everyday events in their world.

Even well-structured problems involve common sense. Consider, for example, the hobbits and orcs problem described earlier. Three hobbits and three orcs arrive at a riverbank with a small boat that will hold two individuals. They want to cross the river, but if the orcs ever outnumber the hobbits on one side of the river, they will eat the hobbits. How do they get across?

There are formal ways to solve this problem. For instance, Saul Amarel has described this problem as a state space consisting of 32 states, two of which are unreachable. He wrote out the solution using a notation consisting of three numbers, representing the number of hobbits, orcs, and boats on the first bank, respectively. The other bank is fully specified once we know the first bank. If the boat is on the side of the first bank, then it cannot be on the side of the second bank, and vice versa. Here is his solution:

$331 \rightarrow 310 \rightarrow 321 \rightarrow 300 \rightarrow 311 \rightarrow 110 \rightarrow 221 \rightarrow 020 \rightarrow 031 \rightarrow 010 \rightarrow 021 \rightarrow 000$

What this notation hides, however, is the commonsense reasoning that goes into understanding the problem from its description. How do we know that Amarel's representation is a fair representation of the problem? For example, how do we know that the presence of the boat is essential to understanding the problem? If it were not, then we would have a very different kind of problem. The hobbits and orcs would simply ford the stream and go on. Problem solved. Recall the problem of finding a pair of brown socks and black socks, the ratio of the two colors was not relevant in that problem, even though most people assume that it is relevant.

In the hobbits and orcs problem we assume that the stream is crossable only with the boat, and it is common sense that supposedly tells us that. We assume that neither hobbits nor orcs can swim across the stream, but that is not stated in the description. We do not assume that the problem is insoluble unless we know what color hat each individual was wearing or unless we know what color the boat was. Common sense tells us that the color is irrelevant. We do not ask if there are oars by which to row the boat. We assume that banks on either side of the river are not so high that it would be impossible to climb or fatal to jump from them into the boat.

The nature of the problem changes dramatically if there is an island in the middle of the river or if every creature has to leave the boat when it reaches shore. The problem also changes if instead of three hobbits and three orcs, we say that four of each type appear on the shore or that more orcs arrive after they have started to cross.

In fact, there are many things that we assume and a huge, perhaps infinite, number of things that we don't consider as relevant. How we translate the words of the problem description into a representation that we can use to solve the problem is a critical issue that is often hidden by the fact that common sense is implicitly used when people solve problems but cannot be assumed when dealing with computers. By the time an engineer has created a representation of a problem, she has used common sense to decide just what features are (potentially) important to the situation and what the relation is between these features and the elements of the representation.

Common sense is essential to creating problem representations and essential to computational intelligence. Even when solving formal problems, how we construct solutions to those problems already depends on commonsense notions about the structure of the problem and the factors that are relevant to it.

Representing Common Sense

Common sense suffers from two so far insurmountable problems. What exactly needs to be represented to capture common sense, and how should that information be represented? Artificial general intelligence is simply not possible without an effective means of representing and exploiting common sense.

Some investigators argue that common sense needs to be represented as a set of facts, for example, organized into a tree. This approach is most consistent with the idea of intelligence being symbol processing. But, as we have been discussing in the context of similarity and categorization, it is unlikely that we can come up with a fixed list of those facts that need to be represented or that we can organize them in a meaningful way.

Instead, common sense involves a kind of logic that is inconsistent with traversing a tree structure like that used to play go or chess. Tree structures and similar forms of deductive logic are "monotonic," meaning, roughly, learning a new piece of information can never reduce the set of what is known. Adding new information, in monotonic logic, always increases the set of facts that are known; it never can contradict what was formerly believed. In fact, the very notion of belief as separate from fact is foreign to monotonic logic.

Common sense requires nonmonotonic reasoning. If you learn that Tweety is a bird, you infer the fact that Tweety can fly. But if you later learn that Tweety is the kind of bird called an ostrich, then you have to revise this fact and then recognize that Tweety cannot fly. Nonmonotonic logic is said to be "defeasible." Beliefs are tentatively held and are subject to revision when more information is gained.

Medical diagnosis is a kind of nonmonotonic reasoning. Although physicians work hard to make medical diagnosis as consistent, systematic, and as logical as they can, ultimately, any diagnosis is an inference from the available evidence and can be contradicted by subsequent information. Reasoning by default, where we believe something to be true until we find out otherwise, is another familiar kind of nonmonotonic reasoning.

Any of the facts that you know, and which you use for problem solving, could be wrong. It is very difficult to reason systematically from premises to conclusions when the facts that those premises are based on could be wrong. Formal, traditional kinds of logic become inconsistent under these circumstances.

Instead, commonsense reasoning is more flexible than traditional formal logic. People can jump to conclusions that cannot be justified. The work of Kahneman and Tversky, as we discussed earlier, shows that people are not consistent in their judgments. Concepts such as bounded rationality also play a role. Herbert Simon, one of the participants at the 1956 Dartmouth Conference on artificial intelligence, argued that people have limited ability to process information. Their decisions and judgments are not fully reasoned but are bounded by the difficulty of the problem, by their own cognitive capacity, and by the time available to reason.

Human common sense reasoning also suffers from a number of so-called cognitive biases. Some of these were discussed earlier in the context of Kahneman and Tversky's work. One of these biases is "confirmation bias." People find it easier to understand information that is consistent with their beliefs than information that challenges them. They tend to look for information that confirms their beliefs, even when they think that they are looking to evaluate those beliefs.

For example, Peter Wason gave people a card task that was designed to investigate how they evaluated hypotheses. He presented four cards, one had an A, one had a 3, one had a B, and one had a 4 showing. The people were asked to evaluate the following hypothesis: If a card has a vowel on one side, then it has an even number on the other. Which cards would you choose?

A 3 B 4

Less than one in four people get this right. Most people pick the first card. If it turned out to have an odd number on its other side, then the hypothesis would be wrong. Many people also choose the last card. They expect that it will have a vowel on the back, but in fact, it does not matter what is on the back of that card. The hypothesis does not say that all even numbers have to have vowels. So whether the back of the 4-card had a vowel or not would not affect the hypothesis. The B-card is also irrelevant because the rule says nothing about nonvowels. It could have any number on its back and not change the truth of the hypothesis. The 3-card, on the other hand is critical. If it has a vowel on the back, then that would make the hypothesis false. It would be a card with a vowel on side and an odd number on the other. Most people choose the cards that confirm the hypothesis (A and 4) rather than the two cards that challenge it (A and 3).

To be fair, people are more likely to get it correct if the task involves more realistic situations—for example, if the rule is, if a person is drinking beer, then the person must be over 21 years old. Their choices are:

Beer Soda 18 25

They realize that they need to know the age of the person drinking the beer, and they need to know the drink of the 18-year-old to know if this rule is correct. But even in realistic situations people tend to show confirmation bias, looking for information that favors their beliefs rather than information that would challenge them. They interpret ambiguous evidence in ways that support their position. They resist changing their beliefs in the face of contrary evidence. The effect is stronger for highly emotional issues and for deeply held beliefs.

The overconfidence effect is another cognitive bias. It is the prevalent belief that we are each above average in some specific features (also called the "Lake Wobegon effect" after Garrison Keillor's stories about cycthe children in the fictional community of Lake Wobegon all being above average). The overconfidence effect is shown when a person's belief in the accuracy of his or her own judgment is higher than the objective accuracy of those judgments. Most people, for example, think that they are better drivers than others. They have an illusion of control, that they are less likely to get in an accident if they are behind the wheel than if they are riding as a passenger. They think that they are more expert than their peers. These beliefs persist despite evidence to the contrary.

A related bias is can be called the "schlemiel/schlimazel effect." If a person is carrying a bowl of soup and spills it on another person, the old story goes, the spiller attributes the accident to some external force—maybe the floor was slippery or something made him trip. The person on whom the soup was spilled tends to attribute the accident to the clumsiness of the spiller. The schlemiel is the idiot; the schlimazel is the victim of bad luck. They may be the same person, depending on who is doing the judging. More formally, this effect is called the "fundamental attribution error."

It is easy to find examples of how human common sense does not conform to formal models of reasoning. It is easy to make people seem stupid. But what we do not know is the role that those same processes play in human intelligence. Their nonconformance with formal reasoning may be precisely why these distortions exist. The limits to human rationality studied by Kahneman and Tversky and by Wason, for example, may be rationality bugs. They may be vestigial remnants of poor brain evolution, or they could be among the very features that make people intelligent. If the mind does not have the time or resources to reason fully, are there shortcuts that mostly work but can sometimes go awry? Recall, for instance, the availability heuristic discussed in chapter 2.

Progress in computational intelligence on such problems as playing go has been achieved, at least in part, by the clever development of heuristics that are incomplete, imperfect, but can be executed in a timely way with a reasonable chance of success. There may be a lot more to be learned from the apparent cognitive biases that play such a substantial role in commonsense reasoning as heuristics for computational intelligence.

Common sense allows us to know that we do not make salads from cotton shirts. If we see a six-foot-tall person holding a two-foot-tall person we do not need to ask which one is the father and which the son. We know what the word "it" refers to if we read a sentence like "I stuck a pin into a radish and afterward it had a hole in it."

Few, if any, current computer programs make effective use of commonsense knowledge. Douglas Lenat, in 1984, started a collection of organized facts called CYC, described in chapter 3. The objective of the project was to capture what it means to have commonsense knowledge. For example, CYC represents the facts that "Every tree is a plant" and "Plants die eventually." These facts allow it to reason that the large apple tree in the backyard will eventually die. CYC's facts were originally hand engineered by having someone actually write down each one. More recently, it has come to use forms of machine learning to augment those handwritten facts. Other programs, such as DBpedia, have been developed to extract knowledge from text, such as from Wikipedia.

CYC also includes an inference engine that allows it to perform logical reasoning based on the facts and relations it contains. The Cleveland Clinic has used CYC to support a system of medical information. Users can ask questions in English. The system then translates those queries for CYC's inference engine, which then tries to derive a meaningful response, based on common sense, medical knowledge, and information about understanding human question patterns.

If commonsense categories could be stored in a taxonomy, reasoning about these categories would be easy. A taxonomy is a treelike collection of categories where lower level categories are a subset of those in the upper levels. For example, the category "animals" might include subcategories "dogs," "cats," "elephants," and "tigers." The category "machines" might include "cars," "trucks," and "computers."

With a taxonomy, it is straightforward to reason about categories and subcategories. If we know that an animal breathes, then we can also know that a cat breathes because cats are a subcategory of animals. But most commonsense categories are not so well structured as a taxonomy would imply. A taxonomy supports indefeasible logic, but common sense is defeasible.

Galileo, for example, is a member of a large and indefinite number of categories, "resident of Pisa," "scientist," "victim of religious persecution," "currently dead person," "historical figure." These overlapping categories make reasoning difficult, and because there is really no limit to the number of categories any particular person or object may be in, categorical reasoning must depend on knowledge that is outside of the taxonomy to select the categories even to reason about.

Furthermore, categories are often ill-defined. What annual income qualifies a person as being "rich"? For most people, it is an income that is greater than their own. Even people with high incomes tend to think of themselves as "well-off." Only people with more money than they have should be categorized as rich. What qualities allow a man to be categorized as "handsome"?

Commonsense knowledge may be important to how humans function in the real world, but so far, it has been difficult to codify this information so that it can support computer intelligence. At this point, I do not believe that we have a good way to systematize commonsense knowledge or even represent it. Even if the "facts" can change, how can we represent their

Figure 9
The moth is inside the circle.

current state in a way that is useful? What facts need to be represented? Are there limits to what gets represented? Is the flexibility of human categorization a feature that is essential to intelligent reasoning or a bug that limits it? I think that these are among the questions we will have to address to get on the path to understanding and creating general intelligence. These are problems that need to be solved at some point.

Resources

Clark, A. (1998). Magic words: How language augments human computation. doi:10.1017/CBO9780511597909.011; http://www.nyu.edu/gsas/dept/philo/courses/concepts/magicwords.html

Barsalou, L. W. (1983). Ad hoc categories. *Memory & Cognition, 11*(3), 211–227.

Behrmann, M., & Haimson, C. (1999). The cognitive neuroscience of visual attention. *Current Opinion in Neurobiology, 9*(2), 158–163.

Dean, T., Corrado, G. S., & Shlens, J. (2012). Three controversial hypotheses concerning computation in the primate cortex. ftp://ftp.cs.brown.edu/pub/techreports/12/cs12-01.pdf

Elderkin, B. (2018). Will we ever be able to upload a mind to a new body? Gizmodo. https://gizmodo.com/will-we-ever-be-able-to-upload-a-mind-to-a-new-body-1822622161

Fischetti, M. (2011). IBM simulates 4.5 percent of the human brain, and all of the cat brain. *Scientific American.* https://www.scientificamerican.com/article/graphic-science-ibm-simulates-4-percent-human-brain-all-of-cat-brain/

Gentner, D. (2003). Why we're so smart. In D. Gentner & S. Goldin-Meadow (Eds.), *Language in mind: Advances in the study of language and thought* (pp. 195–235). Cambridge, MA: MIT Press.

Guitchounts, G. (2009). Cortex rewiring. *The Nerve.* 35-41. https://www.bu.edu/mbs/files/2013/01/NerveFall2009-pdf.pdf

Hebb, D. O. (1949). *The organization of behavior: A neuropsychological theory.* New York, NY: Wiley.

Herculano-Houzel, S. (2009). The human brain in numbers: A linearly scaled-up primate brain. *Frontiers in Human Neuroscience.* doi:10.3389/neuro.09.031.2009; https://www.frontiersin.org/articles/10.3389/neuro.09.031.2009/full

Inafuku, J., Lampert, K., Lawson, B. Stehly, S., & Vaccaro, S. (2010). Downloading consciousness. https://cs.stanford.edu/people/eroberts/cs181/projects/2010-11/DownloadingConsciousness/tandr.html

Kamil, A. C., & Bond, A. (2006). Selective attention, priming, and foraging behavior. *Comparative Cognition: Experimental Explorations of Animal Intelligence.* doi:10.1093/acprof:oso/9780195377804.003.0007; https://www.researchgate.net/publication/282230039_Selective_Attention_Priming_and_Foraging_Behavior/figures?lo=1

Kohler, I., & Erismann, T. (1950). Inversion goggles. https://www.youtube.com/watch?v=jKUVpBJalNQ

McCarthy, J. (1989). Artificial intelligence, logic, and formalizing common sense. In R. H. Thomason (Ed.), *Philosophical logic and artificial intelligence* (pp. 161–190). Dordrecht, the Netherlands: Kluwer Academic. http://www-formal.stanford.edu/jmc/index.html

Medin, D. L. (1989). Concepts and conceptual structure. *American Psychologist, 44,* 1469–1481.

Pietrewicz, A. T., & Kamil, A. (1979). Search image formation in the blue jay (*Cyanocitta cristata*). *Science, New Series, 204,* 1332–1333. Papers in Behavior and Biological Sciences. 65. http://digitalcommons.unl.edu/cgi/viewcontent.cgi?article=1065&context=bioscibehavior

Stratton, G. M. (1896, August). *Some preliminary experiments on vision without inversion of the retinal image.* Paper presented at the Third International Congress for

Psychology, Munich, Germany. http://www.cns.nyu.edu/~nava/courses/psych_and_brain/pdfs/Stratton_1896.pdf

Tinbergen, L. (1960). The natural control of insects on pinewoods: I. Factors influencing the intensity of predation by songbirds. *Archives Néerlandaises de Zoologie, 13,* 265–343.

Wason, P. C. (1966). Reasoning. In B. M. Foss (Ed.), *New horizons in psychology* (pp. 135–151). Harmondsworth, UK: Penguin.

8 Expertise

Psychological investigations of intelligence have focused largely on individual differences among people in their test performance. How people solve problems is more centrally pertinent to developing artificial general intelligence, but perhaps even more central is the process by which people gain expertise. Expertise is more general than the ability to solve individual problems. Expertise is the capability to solve multiple kinds of problems, sometimes even those that have not been seen before.

The people we commonly recognize as intelligent are typically those who have achieved a certain level of success in a particular field. We call these people experts, mavens, prodigies, if they are young, or geniuses. Their success usually extends to a broad area of expertise, not just solving a single problem. As a result, the means by which they achieve this expertise could be extremely informative.

In this chapter we will consider the differences between novices and experts, how human experts gain their expertise, and ask how that expertise may be important to the creation of artificial general intelligence. There seem to be important differences between experts and nonexperts. These differences may be critical to understanding intelligence.

Novices seem to depend more on formal rules to guide them in their performance than experts do. These are the kind of rules that serve as the basis of artificial intelligence. Experts rely to a greater degree on what might be called intuition. Expert go players, for example, describe certain moves as being more aesthetically pleasing than others. That these aesthetic moves are also more likely to be successful is at least part of the reason for these players' success. Intuition in other areas is probably also associated with pattern judgments in experts as opposed to articulated rules in novices.

Expertise may, at least in some cases, depend on specific talents. Was Mozart an expert musician because he was born with some specific talent? The idea of a prodigy, for example, would seem to emphasize the idea that a person has expertise far beyond what others could have acquired through practice at a similar age.

The ideas of talent and expertise are not well distinguished. If a girl is tall, she may be seen as having a talent for playing basketball. If she has the physical stature needed and the interest in the game, she may eventually achieve a certain amount of expertise in the game. As she plays, she finds basketball to be personally satisfying, so she plays and practices more. Does she have talent for basketball, or does she simply have the machinery, if I may be crude, and the willingness to practice? We do not have a good definition of talent that is separated from certain kinds of practice. K. Anders Ericsson has written extensively about the role of talent and practice in developing expertise. We will talk about his work again later in this chapter.

Experts know many things that novices do not know. Expertise consists at least in part of having a deep knowledge of some topic. Chess experts, for example, know a lot about chess and use that knowledge to play it effectively.

In one experiment, a chess player was shown a board holding about 25 chess pieces for 5 to 10 seconds. If the placement of these pieces was taken from an actual game, a chess master can reproduce the position of the pieces with about 90% accuracy. A novice player can usually manage to correctly place 5 or 6 pieces. If, on the other hand, the pieces are placed in random positions, then the chess master falls to the same poor level of performance as the novice. The expert views the pieces in terms of attack, defense, and other structures. Some of these structures have names, such as fianchetto, which is to place a bishop on the long diagonal of the board. Making this move early in a game allows the player to control a large swath of the board.

The difference between random chess positions and sensible positions drawn from a game show that chess masters' performance is not just due to their having better memories (which they may not have) but instead comes from tapping into specialized knowledge that they have about the game. Superior memory comes from being an expert rather than expertise coming from having good memory.

When reproducing the sensible positions of the chess pieces, the experts placed them in groups of about 3 to 5 pieces at a time. Each of these groups is, apparently, a familiar arrangement of what could be encountered in a game. These groups could be said to constitute memory "chunks" similar to those mentioned earlier in the context of remembering long numbers. The experts' "intuition" consists of their memory for these patterns. A random arrangement of pieces, presumably, has few of these recognizable patterns in it and so is difficult to remember. Further, recognition of one of these groups may also lead the expert to think of effective strategies for moving the game forward from these learned positions, which may facilitate memory for other positions. Experts' memory is enhanced by having knowledge of specific strategic arrangements of pieces.

It may not be a surprise that experts have more knowledge than novices, but the difference is not only in amount but in the kind of knowledge. Experts know the vocabulary of their domain (for instance, expert chess players are likely to know the term "fianchettoed bishops"), to be sure, but they also organize their knowledge in different, more effective ways. Chessboard control is more abstract than the specific configuration of the pieces. It may be possible to control the same space on the board with different configurations of pieces, for example. Some of these configurations may be reachable within a few moves of the current position, but others may not be. The fact that they are all somehow equivalent, though, means that the expert player has more opportunity to find one of them and to use it to effectively control the game.

Michelene Chi, Paul Feltovich, and Robert Glaser (1981) investigated how experts differed from novices in solving physics problems. They found that experts and novices categorized problems differently. The two groups represented the problems in different ways. The experts were guided more by physics principles, and the novices categorized problems based on the surface features of the problem. "Surface features" means things like the objects mentioned in the problem description, such as a spring, the exact physics term mentioned, or the relations among the mentioned objects, such as the presence of a block on an inclined plane.

The experts, on the other hand, did not show any particular affinity for the surface features. They did not necessarily group together problems that shared descriptive words or problems with similar looking diagrams. Rather, they categorized the problems using such principles as the conservation of energy law or Newton's second law.

For example, an expert might represent a bar mentioned in a physics problem as a lever. Once represented as a lever, a physics expert has a variety of approaches that can be applied to levers. The novice, in contrast, may not notice that the bar can be abstracted as a lever, and even if the novice did notice this, he might not have the knowledge of approaches that are appropriate to generic lever problems.

In general, the experts used deeper, more abstract, representations of problems. These abstract representations, once identified, make solving the problem easier. Presumably, the experience that made them experts also taught them how to extrapolate from the surface features to the abstract physics principles.

Experts and novices both start with a presentation of the surface features of the problem. Eventually, if they are to solve the problem, they need to come up with one or more mathematical expressions that will let them solve it. The experts are aided in this by learning that a whole class of problems can be solved in a similar way, using similar equations. They need to map the surface features of the problem to these abstract features so they can know how to specify the variables in the abstract representations.

Chi and her colleagues did another experiment with several problems, each employing the same surface structure (for example, weights and pulleys), corresponded to different physics principles. Again, the novices categorized the problems in terms like "rotation," "mass," or "spring." The experts classified them in terms like "conservation of energy" or "conservation of linear and angular momentum." When novices did categorize a problem as an energy problem, for example, the word "energy" appeared in the problem description, even though the problem's underlying physical principles involve conservation of momentum.

Focusing too heavily on the surface features of a problem can lead to "mental dazzle." Children who are capable of solving simple addition problems, for example, may have problems if dollar signs are placed in front of the numbers. Many business people, I have found, find it much easier to understand percentages, such as 73%, than decimal fractions, like 0.73. The dollar sign, the percent sign, and the decimal point are basically irrelevant to understanding the problem but can still be disruptive when the person focuses on the surface structure of the problem rather than on the abstract numerical values involved.

Another significant factor in expertise is the ability to quickly and accurately pick out the relevant parts of a situation. The chess experts reconstructing the positions of chess pieces seemingly had ready-made representations for sensible groups of positions. As we discussed in the preceding chapter, this is another example of a top-down influence on how experts perceive problems.

A few years before Chase and Simon's (1973) study on reconstructing chess positions, Adriaan de Groot (1965) had argued that the main advantage of chess experts is their ability to recognize large numbers of chess positions and the effective moves that follow from these positions. Armed with this information, they would not have to consider all possible moves, only the ones that followed from the positions that they recognized. This view suggests that they might be stymied, then, if their opponent could come up with a new configuration of pieces that was outside of their experience. This is what happened when AlphaGo, the go-playing computer system, made a move that Lee Sedol had never seen or when Lee made a move that AlphaGo had never seen.

The idea that experts are better at selecting potential moves is in stark contrast to an alternative hypothesis that says that experts can process more branches in the tree of potential moves. According to this latter hypothesis, experts evaluate more steps to find the optimum move. If I move this piece to here, then my opponent could make one of several moves, after which, I could make one of several moves, and so on. Going very far along these branches is difficult because of the large number of available combinations of moves and countermoves. Computers can keep track of these trees, at least to some "depth," but humans are limited in the distance they can go. Human chess experts may go a few more steps along the tree than novices do, but the real advantage comes from recognizing the more valuable moves and focusing on them.

Like Samuel's checkers-playing program, chess experts learn from the games that they have played as well as from the games that have been published. Having organized the game position into well-known patterns, they can focus on the moves that have been successful in past games that showed that pattern. There may still be a lot of potential moves to consider, but even these will be way fewer than considering all possible moves.

The same chunks (organized groups of pieces) may occur at several different positions on the board, and the same response strategy may be appropriate no matter where the chunk appears. Rather than representing the entire board with all of its potential positions, the experts may be able to reduce their memory load by representing chunks and moves rather than by remembering positions and moves.

In support of this idea, Heather Sheridan and Eyal Reingold (2014) used eye-tracking technology to determine that experts were better at identifying the most important chunks on the board and at picking the highest quality moves in the context of those chunks. Sheridan and Reingold followed the gaze and move selection of a chess expert and some novices while these players were solving specifically constructed chess problems. They then evaluated the quality of the players' moves using an advanced chess program that was able to more exhaustively evaluate many more possible choices.

Not surprisingly, the experts chose the best move for the circumstance on 93% of the problems, and the novices chose the best move on 52% of the problems. By tracking where the players looked, Sheridan and Reingold found that the experts identified the relevant squares on the board and looked at them sooner than did the novices. Both experts and novices looked more at the relevant than at the irrelevant part of the board, but the novices spent more time examining the irrelevant part than did the experts.

In short, Sheridan and Reingold verified de Groot's prediction that experts not only make the correct move more often but they use their knowledge of previous chess games to be more selective about which moves they evaluate. Like experts in other areas, they apparently treated the chessboard configuration in a more abstract way than merely memorizing the surface configuration of pieces. Rather than treating any move as equally worthy of consideration, the experts were better able to identify a subset of moves corresponding to each situation and preferentially analyze them.

Expertise has also been studied in sports performance. Problems are more difficult to present in a sports study than in a chess study, but assessing the quality of the result is usually still manageable. For example, the ability of novice and expert squash players has been studied by showing participants videos of players in actual games. Experts are better than novices at anticipating where a shot will go after watching a brief video of the play. They need less information (shorter video segments) than novices do. Experts can better anticipate the direction of a shot from early parts of an opponent's actions than novices can.

Expert snooker players do not differ from novices in tests of visual acuity, color vision, depth perception, or eye-hand coordination. Snooker, by the way, is a game very similar to pool or pocket billiards. It was invented in India in the nineteenth century during the British occupation. Like chess players, expert snooker players were better able to recall and recognize pictures depicting normal game situations. Also like chess players, they were no better than novices at remembering random ball placements. The superiority of expert snooker performance, then, appears to be due to the expert players' experience rather than to any inherent difference in their motor or perceptual capabilities.

Similar patterns of expertise have been found in many other domains, including computer programming, history, electronic circuitry, teaching, physics, badminton, medical diagnosis, bridge, and radiology. Basically, wherever investigators have looked, they have found similar patterns of difference between experts and novices in that domain.

When solving physics problems, experts frequently mentioned the physical principles or laws that they would apply in solving the problem. Novices tended to talk about the equations that they would use. Experts might draw simple diagrams of their problems where novices would focus on plugging numbers into their equations. Knowing more appears to entail having more conceptual units available in memory and more relations and more meaningful relations among these units.

The problems that were traditionally studied within the context of human problem solving included path problems like the Towers of Hanoi, hobbits and orcs, and others. These problems were easy to study because they did not require any specific outside knowledge. Experimental participants are readily available. Expertise, on the other hand, does require knowledge, which may take years to develop and may not be widespread. As a result, expertise is more difficult to study than other forms of problem solving, and it is a challenge to recruit participants.

Studies of expertise, unlike other forms of problem solving, seek to understand just what knowledge the expert brings to a problem and how that knowledge is deployed in its solution. The expert solver must either select a framework for solving the problem or perhaps even invent one. Experts put the problem in context and use that context as part of the process of solving it. Experts do not just work to find a path through a well-structured problem but must supply the path as well.

The development of expertise is the development of knowledge and of the tools and strategies to employ it to solve expert-type problems. In chess, we have a good idea that some of that knowledge consists of patterns of pieces and their strategic role in moving the game toward a conclusion. Other expert tasks would seem to employ similar pattern representation and recognition capabilities. Expertise seems to depend also on the development of relatively abstract representations.

Despite the superior performance of experts over novices, experts may not always be able to describe how they solve problems. Process description is relatively easy when the person is following specific rules, but when an expert's performance depends on his or her ability to perceive patterns, these patterns may not be simple and the process of applying them may be difficult to articulate. That is probably why go players describe their move choices in aesthetic terms and why we think that experts depend more on intuition. Intuition may be nothing more than basing decisions on patterns that are difficult to describe explicitly.

To summarize:

- Experts have knowledge that novices do not have.
- Experts recognize and act on features and patterns that are not used by novices.
- Experts organize their knowledge differently from novices.
- Experts describe problems in more abstract ways than novices do.
- Expert knowledge is more global than novices' knowledge is.
- Experts can retrieve information with less explicit effort than novices can.
- Experts have access to more strategies than novices have.
- Experts appear to have better intuition about problems than novices do.

Expertise is an essential part of what we mean by general intelligence. The mechanisms and skills that constitute expertise, however, seem to be substantially different from those used to study the kinds of path problems that have been the research target of human problem solving as well as the kind of narrow artificial intelligence that is characteristic of current work in computational intelligence.

Source of Expertise

Rather than attributing exceptional performance to perhaps inherited talent, recent research on experts has found that superior performance is associated with specific forms of practice engaged in for a suitable amount of time. The type and duration of this practice seems to be extremely important in determining just who can achieve this exceptional level of performance.

Even with practice, a person's biology may still have an effect. It is much less likely that a short person of 5 feet 3 inches could be a professional basketball player, let alone one who demonstrates superior levels of performance, than someone who is 6 feet 8 inches tall, but it does happen. The shortest player ever to play in the NBA was Muggsy Bogues, who was 5 feet 3 inches tall. He played in the NBA for 14 years. In his years at Charlotte, he became the Hornets' career leader in minutes played (19,768), assists (5,557), steals (1,067), turnovers (1,118), and assists per 48 minutes (13.5).

Tall people typically have an advantage as basketball players, but the Bogues experience suggests that it is not all there is. Linemen in the National Football League are all very large, and there is not much variation among them in terms of body mass. In the 2016 season, the average NFL lineman weighed 315 pounds. Yet some of them are much better players than others. The average body weight of an NFL lineman had no observable effect on the number of yards that team gained per game. With this narrow weight range (304–327 pounds), better players would have to differ from poorer players on some dimension other than weight. The most important way that they differ might be in how they trained and practiced.

IQ and Expertise

As it turns out, there is only a weak relation between IQ, at least as measured using IQ tests, and ultimate achievement in many domains, including go, chess, and music. Scientists, engineers, and medical doctors who have completed the required education show practically no relation between measures of professional success and IQ. That caveat may be important. Medical school admissions typically depend on scoring well on the MCAT or other exams. Mostly only high-scoring students make it into and through medical school, so that by the time that they graduate there is not much range left to account for ultimate professional success. Low-scoring medical students and puny linemen do not get admitted to medical school or get drafted by the NFL. Statisticians call this phenomenon "restriction of range." The contribution of IQ may, in other words, already have been parsed out before these students finished their education. Any remaining difference, then, would have to be due to factors other than the intelligence measured by the admission test. With a few exceptions, such as Bogues, most basketball players are tall, so variations in height are an unreliable predictor of how well a player will play.

In general, many kinds of ability or aptitude tests are good at predicting how well a person does when starting a new educational experience or a new career, but they are generally poor predictors of a person's ultimate success. Even after adjusting for restriction of range, the longer a person spends in a job, the less that person's performance is predicted by ability and aptitude tests. Even the SAT test, which most colleges require and on which they make their admission decisions, is only a weak predictor of the student's fourth-year grade point average.

It's worth emphasizing that IQ and related tests may be good predictors of how rapidly, if you will, people learn introductory material, but they seem poor at predicting ultimate attainment. Other variables such as practice seem to make much more of a difference. General ability, in short, seems to reflect the general ability to learn basics, but not the general ability to achieve a level of expertise. There may be a minimal level of intelligence necessary for intellectual achievement, but given that the minimum is exceeded, other factors typically better account for the measured outcome.

For example, an analysis of the relationship between intelligence and chess ratings over 19 studies by Alexander Burgoyne and his colleagues (2016) found a moderate relationship, particularly for early career players and those at lower skill levels. The relationship was much weaker at high skill levels. As with other kinds of expert performance, the tests seem to be predictive of early success, but not of ultimate success. An interesting part of this chess finding is that there is no explicit selection by aptitude test for becoming a chess player as there is for medical school, so restriction of range may have a smaller role in explaining this lack of correlation.

Fluid and Crystallized Intelligence

Psychologists often distinguish two kinds of intelligence. One is called "fluid intelligence," which includes problem-solving and abstract reasoning. The second is called "crystallized intelligence," which is much more knowledge based. Generally, as we age, the role of crystallized intelligence grows as we acquire more knowledge.

Fluid intelligence is associated with induction—the process of inferring rules from examples and the capacity to form concepts; visualization—the ability to construct images; quantitative reasoning; and ideational fluency—the ability to generate ideas, for example, in brainstorming. Crystallized intelligence is associated with verbal ability, reading comprehension, sequential reasoning, and knowledge of general information. Put crudely, fluid intelligence is what you need to figure things out. Crystallized intelligence is what you use to apply knowledge that you have.

Fluid intelligence tends to increase until young adulthood and then declines. Crystallized intelligence increases gradually and remains stable until sometime around age 65, when it begins to decline.

This distinction between fluid and crystallized intelligence is consistent with the findings of Burgoyne and his colleagues. At early stages of a chess player's development, the ability to reason quickly is important, but with increasing experience, knowledge patterns seem to play a larger role. As a chess player learns more patterns and appropriate responses to them, the player does not need to figure out the response; she just needs to retrieve it from memory.

People with high fluid intelligence tend to process information more quickly, have moderately greater memory spans, and use more sophisticated strategies, compared with lower-scoring individuals. These capabilities let them solve problems faster and more accurately than less intelligent individuals, but these benefits disappear with age or experience.

As people age past early adulthood, the speed of their performance typically declines, but their knowledge generally increases. It does not seem to make much sense to say that healthy middle-aged people are less intelligent than their younger selves. Rather, people tend to substitute knowledge for rapid analysis. By analogy, we might say that mature people store solutions rather than compute them.

The ideas of fluid and crystallized intelligence started with intelligence tests. The statistical approach to analyzing the various parts of intelligence tests broke these parts into components, called factors (see chapter 2). The idea is that any one subtest of intelligence would be associated with one or more of these factors. Factor analysis, as the statistical process is called, looks at the correlation patterns among the subtests and figures out the underlying statistical measure or factor that would best represent the correlations. Raymond Cattell identified groups of tasks that were associated with crystallized intelligence or with fluid intelligence. Subsequent studies looked at how these two factors changed over time as people aged. The same factor analysis also identifies a general factor (among others), which is called general intelligence, or "g." It takes note of the fact that there is some correlation among all of the various subtests. These tests and factors are discussed in more detail in chapter 2.

The shift from fluid to crystallized intelligence could turn out to be critical to constructing a computational intelligence. Much of the effort in machine learning, particularly in deep neural networks, is analogous to crystallized intelligence. The network learns patterns that can be used to decide on actions. The actions are typically categorization judgments, but they could be other kinds of things as well. Improving the raw computational power of computers is somewhat related to fluid intelligence in that decisions and measurements can be conducted more quickly and efficiently. However, there is little computational work on "figuring things out." Rather, machine learning seems limited to selecting among a potentially large set of given alternatives, which would seem to be more closely allied with crystallized intelligence.

On the other hand, although most computational intelligence depends strongly on stored knowledge patterns, it is unclear how the kinds of knowledge representations that allow expert chess players to focus on the kind of move that will be most valuable can be replicated in computational intelligence. How do chess players organize chess pieces into groups? Which pieces do they include? How do they decide which of these groups is more important than others? The primary means that computer models have for organizing parts into groups is co-occurrence. Pieces that occur together in multiple games, for example, could be organized into a chunk. At present there is no means for using more abstract properties of the group, such as, its ability to control a part of the board, to organize them into a chunk. These are not insurmountable problems for machine intelligence, but they are still largely open questions.

The Acquisition of Expertise

Intelligence tests were originally designed to identify and measure individual differences to make appropriate decisions about student placements. The tests were intended to identify the indicators that correlated with success in school. The testing movement depended on the assumption, sometimes explicitly made, that these tests measured something fundamental, immutable, and biological about a person, something that was not itself changed by education, but that indicated a capacity for education. Success, or even superior performance, is, in this view, determined by basic unchangeable endowments.

Charles Darwin's cousin, Sir Francis Galton, is often recognized for developing this view in the nineteenth century. Galton examined the relatives of famous intellectuals and found that the probability of "genius" declined as the distance of the relationship increased. Brothers of geniuses were more likely to also be geniuses than were cousins of geniuses. He also did some twin studies, finding that twins raised apart were more similar than non-twins raised together. His strong belief in the inherent biological basis of intelligence and his ideas about their heritability led him unfortunately to eugenics, the idea that it is possible to selectively breed for intelligence among humans.

On the biological view, intelligence tests are used to select students with the right set of innate capacities to allow them to become experts in their chosen domains. The evidence for such innate abilities is scant. Assuming that intelligence tests assess these innate factors, we find performance on age-appropriate tests of intelligence tend to be relatively stable over time. Intelligence and related tests predict how well a student will do in school, and similar tests predict how a person will perform on a new job, but they are not particularly good, as we have seen, at predicting a person's ultimate level of attainment.

In contrast to this traditional view, the only two genetic characteristics for which we have definite evidence of their influence on highly skilled performance are height and body size. Above-average height is an advantage in basketball, and below-average height is an advantage in achieving elite performance in gymnastics.

Experts and elite athletes typically need to spend about 10 years engaged in perfecting their craft to achieve elite levels of performance.

The scientific investigation of expertise depends to a large extent on having repeatable measures of performance. The focus on athletics, chess, music, and similar activities is not because these are somehow special activities but because performance on them can be objectively measured. People can have reputations as elite performers without actually having to accomplish anything elite, for example, if people have expectations about how important they are. An expert's reputation can exceed any objective assessment of that performance (consider the overconfidence effect of the preceding chapter). Anecdotes are typically not a sufficient basis for the development of scientific theories. Ostensibly successful financial advisors, for instance, are often found to be no better than others at picking stocks. Chess, on the other hand, has an objective ranking system (the Elo rating system), which involves matches entered and won and the quality of opponents played.

The focus on objective, reproducible measures provides a framework for assessing theories of expertise. In this context, we find that the level of expertise increases gradually over time. There is little evidence of a reliable "aha" moment when a sudden insight transforms a person from a novice to an expert. Even child prodigies, when measured in a consistent way against adult standards, show evidence of a long, gradual path to improvement.

The age at which experts achieve their peak performance tends to be in their 20s for many sports, and in their 30s or 40s for less intense sports and for the arts and sciences. Even the most "gifted" performers need about 10 years of intense practice before they reach high levels of achievement, such as being able to compete at an elite international level in sports.

Just playing golf for 10 years is not enough to ensure a high level of success in the game. Rather, a specific kind of effort seems to be required. Anders Ericsson has called this kind of practice "deliberate practice." It is not intrinsically fun. It is deliberate work intended to improve specific parts of the person's performance on a well-defined task, usually measurable. It involves detailed immediate feedback as to the success of the performance. It involves significant repetition of the same or similar tasks over time.

Tiger Woods, for example, practices a putting drill where he places two golf tees in the ground, separated by about the length of the head of his putter, about three to four feet from a hole. He would place a golf ball between the two tees and putt one-handed and two-handed until he had sunk 100 balls in a row.

All expert musicians practice, but elite musicians spend more time than others on solitary practice. By the time they were 20, the best expert musicians had done over 10,000 hours of practice, compared to a group of less accomplished musicians, who had spent between 5,000 and 7,500 hours in practice, or amateur musicians, who had spent around 2,000 hours in practice.

When the expert musicians practiced, they concentrated on specific aspects of the music performance as directed by their music teachers. The best expert musicians practiced like this for about four hours every day, including weekends.

Improved practice methods have resulted in major changes in athletic performance over the years. Olympic gold medalists in sports that can be objectively measured (for example, runners measured by running times) have improved 30% to 50% from the beginning of the modern Olympic games until recently. Some of this improvement is due to the use of better tools to analyze performance imperfections, which, in turn, have led to better, more deliberate practice.

Expert chess players practice by studying published games between the best chess players they can find. They go through these games move by move, predicting what the expert will do. If their prediction differs from the actual move in the game, they attempt to figure out why they chose differently. Serious chess players spend around four hours per day practicing like this.

Elite musical composers also require about 10 years of experience. J. R. Hayes (1981) analyzed the productivity of 76 composers for whom he could find sufficient information about when they began their intensive study of music. Only three of them produced significant compositions in less than 10 years after starting their intense music study (Satie in year 8 Shostakovich and Paganini in year 9. Most of the significant works by the full set of 76 composers were produced between years 10 and 25 after starting intense instruction.

Wolfgang Amadeus Mozart is arguably the most elite musical composer that the world has known. His talent, like that of other composers, was cultivated, through long practice. Leopold Mozart, Wolfgang's father, was a composer, musician, and music teacher. The younger Mozart was taught music from an early age, particularly composition. He was invited to participate with other highly skilled musicians visiting the Mozart household. Mozart's early concerti were not original compositions but were arrangements of the works of other composers. Like other prodigies, these works may have been outstanding when coming from one so young, but when judged by adult standards, they were not yet sophisticated (Hayes, 1981; Weisberg, 2006).

A historical analysis of the most important scientists and poets of the nineteenth century found that the average age at which the scientists published their first work was 25.2 years. The average age at which these scientists produced their greatest works was 35.4. Poets and authors published their first works at an average of 24.2 years of age and their greatest work at age 34.3 years.

The same 10 years of preparation has been noted in musical performance, mathematics, tennis, swimming, diagnosis of radiographic images, medical diagnosis, and long-distance running. Simon and Chase (1973) got it about right when they argued that 10 years of deep practice is necessary to develop elite levels of performance in a wide variety of domains.

So large amounts of deliberate practice appear to be necessary for the development of elite talent in a wide variety of domains. Intellectual achievements, such as writing poetry, conducting scientific research, or composing music, are extensions of the kind of tasks that we associate with intelligence. It's not clear what these activities have in common with sports and similar kinds of performance (such as dance or musical performance). It is possible that the same need for extended deliberate practice pertains to many other domains, but that it is difficult to scientifically investigate these other domains, for example, because they have less clear-cut criteria for success. In any case, this need for extended practice contrasts with the kind of experience that gives rise to competence. Infants are born without knowing any words, but by the time they are 12-year-old children, they may know more than 50,000 words. They learn words at an average rate of over 10 new words a day (there are about 4,400 days in 12 years). By the time they are 12, they are competent language users, but few could be called elite performers.

Elite intellectual achievement seems to require the same kind of repetitive exposure that modern neural network models and deep learning models seem to require. A child may need only one exposure to learn the meaning of a word or to learn that kids get cotton candy when they go to the zoo but may need 10 years of extensive purposeful practice to become an elite poet writing about cotton candy. Here (https://www.poets.org/poetsorg/poem/cotton-candy) is a poem about cotton candy by Edward Hirsch. You can judge for yourself whether it achieves elite status.

Artificial intelligence research tends to be more concerned with quotidian activities such as reading handwritten characters or driving cars down crowded highways. But there may be something important to learn from those people who learn to perform at an elite level.

An example of how elite performance emerges from the knowing application of simple processes is how baseball outfielders position themselves to catch a batted ball. They do not perform deep mathematical calculations involving parabolic trajectories. They seem to depend on depth perception only during the final stages of the catch. Until the ball is very close, depth cues, such as eye convergence, are unavailable. As objects approach, your eyes angle toward one another, and the angle of this convergence is a cue to how close the object is, but these cues are only available when the object is very close. According to studies by Michael McBeath and his colleagues (1995), outfielders do something much more practical and simple.

McBeath and his colleagues found that outfielders use a basic visual cue to tell them where to run to. They track changes in the image of the ball relative to its background. Examining this research in more detail is instructive because it shows the representation that these players use in solving the problem of catching the ball. They construct a representation that is easy to compute and that requires the least amount of running to catch the ball.

One potential representation that fielders could use would be to compute the parabolic path that the ball will take. This computation is what we use to send a rocket to the moon or to launch an artillery shell. Alternatively, the fielders might have constructed a mental model from their experience with balls and gravity. Starting with an estimate of the ball's speed and direction, the model could predict where the ball will land. Although this mental model was a long-held theory, it requires that the fielder accurately perceive the ball's motion as it comes off the bat and then accurately compute its trajectory, taking into account the ball's spin, wind speed, and air density. Given that the fielder is more than 30 meters (more than 98 feet) from the bat when the ball is hit, accurate estimation of the necessary parameters seems very unlikely.

Two other potential representations do not require the fielder to compute the trajectory of the ball. Instead, these hypotheses suggest that the fielder can anticipate where the ball will land based on continuously updated visual information. According to these hypotheses, the fielder does not have to predict where the ball will land—a gust of wind, for example, would destroy that prediction. Instead, the fielder uses visual cues to guide where to go during the ball's flight. The fielder keeps his eye on the ball.

To a stationary fielder, the ball appears to rise during the initial part of its path and then appears to fall. If the fielder moves, however, his or her motion will affect whether the ball appears to be rising or falling and affect its lateral position relative to the fielder.

According to one theory, the fielder responds to the visual acceleration of ball. If the ball's optical velocity (that is, its visual movement relative to the background) is increasing, then the ball will land behind the fielder. If the ball's optical velocity is decreasing, then the ball will land in front of fielder. By running in such a way that the ball is seen to be neither accelerating nor decelerating, the fielder will be in position to catch the ball. By keeping the apparent (the visual as opposed to the physical) velocity of the ball constant, the fielder will end up in the right place. This hypothesis is called "OAC theory," optical acceleration cancellation, because of its prediction that fielders work to cancel the apparent acceleration of the ball during its flight.

According to a third hypothesis, the fielder runs to keep the apparent visual trajectory of the ball moving in a constant direction relative to the horizon. This hypothesis, called "LOT theory," linear optical trajectory, is simpler than the acceleration theory because it does not require the fielder to detect whether the image of the ball is accelerating or decelerating, just to detect whether it is moving in a consistent direction.

These three accounts attribute different representations to the fielder. The first one argues that the fielder has a detailed model of the ball's trajectory. The other two accounts do not represent the position of the ball at all; they argue that the problem can be solved by responding to the *appearance* of the ball. Of the appearance accounts, I think that the evidence favoring the linear model is stronger than the evidence favoring the acceleration model, but it scarcely matters for our purposes. The important point is that choosing one kind of representation, in this case a visual one, makes an intractable problem, trajectory estimation, much simpler. Dogs cannot compute parabolic trajectories, but many of them can catch balls or Frisbees.

It is not clear whether this simplified representation is invented or is built into the brain somehow. Any predator would have to be able to respond to prey it was trying to capture, so it may be that this process evolved early among predators and is simply exploited by outfielders and dogs. Or it could be that each fielder or predator learns this relationship through experience catching things. In either case, complex problems can often be solved using simpler solutions, which entail simpler representations. This example is illustrative of the kinds of tools that experts can come to apply to complex problems using their crystallized intelligence. And, as it turns out illustrative of the kinds of solutions that have led to current models of computational intelligence.

A potentially complex task turns out to be solvable by a relatively simple algorithm—one that is unconsciously applied. Few, if any, fielders are likely to be able to articulate this linear optical trajectory mechanism, beyond the saying to "keep your eye on the ball." Yet the available evidence suggests that this is, in fact, how it is done. As in successful applications of artificial intelligence to problems like playing go or chess, this solution is not a complicated strategic analysis or deep calculation but a simple solution to an apparently complicated problem.

One of the most important points to take away from our discussion of expertise is that the nature of the representations people use changes over time. As expertise grows, experts come to represent situations in more abstract, more principled ways. They move from surface analyses and rote rule following to new solutions that are based more on analysis of abstract patterns. In the beginning, these abstract patterns may not be apparent, but with extended experience, they come to dominate the expert's thinking process.

Machine learning and computational intelligence have not yet gotten to the point where they can change their representations in quite the way that human experts do. On the one hand, deep neural networks can learn abstractions of their input patterns. But the difference between computing a trajectory and using the appearance of a ball does not seem to be a simple transformation of one to another. The visual approach seems to be a radical reorganization of the problem into a new representation in the same way that changing our conceptualization of chess playing from a psychological matching of wits to one of traversing a tree of possible moves was a radical reorganization. Casting chess as a tree to be traversed made the problem much more tractable than trying to guess at the opponent's psychological motives for each move.

In neither case was the change in representation discovered by a machine learning system. It was designed by a human. Current computational intelligence techniques do not provide any mechanism by which such representational insights can be achieved.

Machines have, of course, been designed to do online tracking of objects. Servo mechanisms, even simple thermostats, adjust their state in response to simple sensory elements. Guided missiles use a mechanism similar to these visual models to track their targets. These mechanisms were designed by engineers, however, not discovered through machine learning. General computational intelligence that mimics human cognition will require mechanisms to discover or invent such solutions at some point.

One of the other things that we can learn from this understanding of human elite performance is that there may really be no substitute for extended experience. If Ericsson and his colleagues are correct, then it could mean that such practice is sufficient even for computers, at least computer systems of a particular type. Endowed with some minimal capacity, a computer with enough labeled examples or similar feedback should be able to become an expert in some domain. AlphaGo played millions of virtual go games. Ericsson claims that what we think of as talent is irrelevant to human accomplishment, so it may also be irrelevant to computer accomplishment. There may be no shortcuts to becoming an expert, but there may be no fundamental barrier either.

At this point, we are still missing a few of the necessary capabilities. There is no guarantee that computational intelligence needs to mimic in detail human intelligence, but so far, these are the only methods that we know of that do implement general intelligence. If these processes can be computationally implemented, we may be able to achieve general computational intelligence.

Resources

Abernethy, B. (2008). Anticipation in squash: Differences in advance cue utilization between expert and novice players. *Journal of Sports Sciences, 8,* 17–34. doi: 10.1080/02640419008732128; http://shapeamerica.tandfonline.com/doi/abs/10.1080/02640419008732128

Abernethy, B., Neal, R. J., & Koning, P. (1994). Visual–perceptual and cognitive differences between expert, intermediate, and novice snooker players. *Applied Cognitive Psychology, 8,* 185–211. doi:10.1002/acp.2350080302; http://onlinelibrary.wiley.com/doi/10.1002/acp.2350080302/abstract

Burgoyne, A. P., Sala, G., Gobet, F., Macnamara, B. N., Campitelli, G., & Hambrick, D. Z. (2016). The relationship between cognitive ability and chess skill: A comprehensive meta-analysis. *Intelligence, 59,* 72–83. doi:10.1016/j.intell.2016.08.002

Chase, W. G., & Simon, H. A. (1973). Perception in chess. *Cognitive Psychology, 4,* 55–81.

Chi, M. T. H., Feltovich, P., & Glaser, R. (1981). Categorization and representation of physics problems by experts and novices. *Cognitive Science, 5,* 121–152. https://pdfs.semanticscholar.org/16ef/4cc3a80ee7ba8f59e0a55b2ef134c31e18b3.pdf

Chi, M. T. H., Glaser, R., & Rees, E. (1982). Expertise in problem solving. In R. Sternberg (Ed.), *Advances in the psychology of human intelligence* (Vol. 1, pp. 7–75). Hillsdale, NJ: Erlbaum. http://chilab.asu.edu/papers/ChiGlaserRees.pdf

de Groot, A. D. (1946). *Het denken van de schaker* [The thought of the chess player]. Amsterdam, the Netherlands: North-Holland. (Updated translation published as *Thought and choice in chess*, The Hague: Mouton, 1965; corrected second edition published in 1978.)

Ericsson, K. A. (2004). Deliberate practice and the acquisition and maintenance of expert performance in medicine and related domains. *Academic Medicine, 79*(10), October Suppl. http://edianas.com/portfolio/proj_EricssonInterview/articles/2004_Academic_Medicine_Vol_10,_S70-S81.pdf

Fink, P. W., Foo, P. S., & Warren, W. H. (2009). Catching fly balls in virtual reality: A critical test of the outfielder problem. *Journal of Vision, 9*(13), 14.1–8. doi:10.1167/9.13.14; https://www.ncbi.nlm.nih.gov/pmc/articles/PMC3816735/

Gobet, F., & Simon, H. A. (1996a). Recall of rapidly presented random chess positions is a function of skill. *Psychonomic Bulletin & Review, 3,* 159–163.

Hayes, J. R. (1981). *The complete problem solver.* Philadelphia, PA: Franklin Institute Press.

Hirsch, E. (2010). Cotton candy. https://www.poets.org/poetsorg/poem/cotton-candy

Larkin, J. H., McDermott, J., Simon, D. P., & Simon, H. A. (1980). Models of competence in solving physics problems. *Cognitive Science, 4,* 317–345. https://www.researchgate.net/profile/John_Mcdermott10/publication/6064271_Expert_and_Novice_Performance_in_Solving_Physics_Problems/links/5489c30f0cf214269f1abb55.pdf

McBeath, M., Shaffer, D., & Kaiser, M. (1995). How baseball outfielders determine where to run to catch fly balls. *Science, 268,* 69–573. doi:10.1126/science.7725104; http://www.bioteach.ubc.ca/TeachingResources/GeneralScience/BaseballPaper.pdf

McPherson, G. E. (2016). *Musical prodigies: Interpretations from psychology, education, musicology, and ethnomusicology.* Oxford, UK: Oxford University Press. https://books.google.com/books?id=3ATnDAAAQBAJ&dq=10+years+practice+musical+composition

National Research Council. (2000). *How people learn: Brain, mind, experience, and school: Expanded edition.* Washington, DC: National Academies Press. doi:10.17226/9853

Chapter 2: How experts differ from novices. https://www.nap.edu/read/9853/chapter/5

Sheridan, H., & Reingold, E. M. (2014). Expert vs. novice differences in the detection of relevant information during a chess game: Evidence from eye movements. *Frontiers in Psychology, 5,* 941.

Simon, H. A., & Chase, W. G. (1973). Skill in chess. *American Scientist, 61,* 394–403. https://digitalcollections.library.cmu.edu/awweb/awarchive?type=file&item=44582

Simons, D. (2012). How experts recall chess positions. http://theinvisiblegorilla.com/blog/2012/02/15/how-experts-recall-chess-positions/

Weisberg, R. W. (2006). *Creativity: Understanding innovation in problem solving, science, invention, and the arts.* Hoboken, NJ: Wiley.

9 Intelligent Hacks and TRICS

To what extent are specialized learning mechanisms necessary to explain general intelligence? Are there certain phenomena that require a special-purpose module, or are all learning problems susceptible to a general learning mechanism? This chapter starts with a critique of connectionist models that were purported to learn linguistic skills in the same way that human children do. Instead, we are reminded of the critical role that representation plays in machine learning. We then turn to contrasting accounts of the potential special-purpose mechanisms that may be necessary and find those wanting. We conclude that problem-specific representations, but not problem-specific learning mechanisms, are required for today's machine learning. More general mechanisms will be needed for general intelligence.

In the preceding chapter, I drew a distinction between the kind of mechanisms that people employ when they learn language and the kind that they employ when they learn to become experts. Almost all people learn language during their early childhood development, typically with little consistent or formal instruction. Experts, on the other hand, require extensive deliberate practice. Is this difference a matter of degree, or is it a difference in the kind of mechanism used in the two learning tasks?

With the capabilities shown by the resurgence of artificial neural networks during the 1980s and early 1990s, the question concerning a need for specialized learning mechanisms became a hot topic among a broad swath of academics. Strong claims were made that basic neural network models could learn language properties that were previously thought to require special learning mechanisms. For example, David Rumelhart and James McClelland (1986) argued that their connectionist neural network models could learn regular and irregular patterns of past-tense formation from experience with no other linguistic knowledge. More, the order in which these characteristics were acquired was very similar to that shown by children. Their network was designed to take as input the present tense of some verb, for instance, "fish," and to output the past tense of that verb, "fished."

During about this same time, comparative psychologists, such as Herbert Terrace, Louis Herman, Sue Savage-Rumbaugh, and Duane Rumbaugh, were looking at the possibility of teaching language to chimps and other animals, including dolphins. The main question that they were hoping to answer was whether language can be learned by a general learning system or whether it depends on the existence of a special language-learning mechanism possessed only by humans. Does the problem of learning language require, in other words, some human-specific representation? If language can be learned by nonhumans and by machines, that would seem to indicate that ordinary learning mechanisms are sufficient. The deeper expectation was that if these linguistic characteristics could be learned by a general-purpose learning system, then it lent confidence to the possibility that these mechanisms would also be sufficient for constructing general intelligence.

The connectionist and comparative claims were disputed by linguists and psycholinguists. These researchers argued that real language was learned only by people and that only people had the capability of learning that language. Noam Chomsky, for example, argued that humans learned language because they were endowed by evolution with a "language-learning organ" as part of the human brain. He claimed that there was no other way to account for the fact that human children learn language at all. The connectionist and comparative psychologists were arguing that experience was sufficient; no particular structures were necessary. The linguists were arguing that the reason people learn language and other animals do not is because people have specific structures in the brain that already represent language. Experience just optimizes.

This question concerning the need for special mechanisms for certain capabilities is central to the idea of creating a general artificial intelligence. Just how many learning mechanisms would such a system need, and what characteristics would these learning mechanisms need to have? On the other hand, the idea of a general learning mechanism that would be sufficient for all tasks would greatly improve the chances of creating an artificial general intelligence.

Noam Chomsky came to prominence in the field that would eventually become cognitive science starting with his devastating review of (B. F.) Fred Skinner's book *Verbal Behavior*. In the book, Skinner tried to lay out a general-purpose mechanism that could account for language learning using basic principles of reinforcement. Skinner's reinforcement learning was similar to today's connectionist model of the same name. It can be summarized by saying that behaviors that are followed by reward occur more often.

Skinner thought that he could identify the variables that control verbal behavior to determine a specific verbal response. In his view, environmental stimuli control which behaviors are "emitted" in their presence. Reinforcement changes the probability of, that is, controls, the verbal behaviors as well as other kinds of behavior. Put simply, people say "red" in the presence of red things because they have been previously rewarded for saying "red" in similar situations.

Chomsky did not have much patience for Skinner's approach. As an example of how Skinner would attempt to explain language, a person might see a piece of music and say "Mozart," because that person has, presumably, been rewarded in the past for saying the name of the composer. But if the person said something else, Skinner would equally explain that utterance by pointing to some other property of the object and talk about how the person *must have been* reinforced in the past for saying that in the presence of that stimulus.

A person could say "leather" in response to a leather chair, presumably in response to its upholstery. Alternatively, the person could say "sit" in the same situation, again presumably because of past reinforcement. Because the reinforcement history of people is complex and mostly unrecorded, there was no independent basis to establish whether these assertions were correct or not.

Chomsky pointed out that Skinner's reasoning was entirely circular. Any utterance could be explained after the fact by pointing to some presumed property of the environment and some presumed reinforcement history. Chomsky claimed that these circular explanations were nothing more than "play-acting at science." We might now recognize that reinforcement learning may, in fact, play a role in the acquisition of words, but without a detailed history, the so-called explanation was empty.

Chomsky argued that many of the patterns seen in human language could not be learned by any known learning mechanism. By that, he appears to mean that it could not be learned by the kind of reinforcement mechanism that Skinner described, or perhaps by imitation. Even in 1959, other forms of learning were known, but Chomsky does not seem to have paid much attention to them. If language properties could not be learned, then they must be innate properties of human brains. If reward and imitation could not generate linguistic patterns, then these patterns must be innate, he argued.

There are several phenomena that Chomsky argued were so rare that children could not have had the opportunity to learn them. Among these are "parasitic pronouns." These are pronouns that can be dropped from a sentence without changing its meaning or grammatical correctness. You can say, "Which article did you file without reading it?" You can say, "Which article did you file without reading?" You can include or omit the pronoun "it" without changing anything significant about the sentence. In contrast, you might say, "John was killed by a rock falling on him," but you would not say, "John was killed by a rock falling on."

Chomsky's argument is also problematic. Chomsky could not marshal any specific evidence that would compel his particular viewpoint. Rather, his argument was focused on absence of a learning mechanism to which he could point and on some of the same kind of circular reasoning he attributed to Skinner.

Even though Skinner could not demonstrate how a phenomenon was learned or Chomsky could not demonstrate the brain mechanism by which it was produced, it could still possibly be learned by some mechanism that neither of them knew. Rummelhart and McClelland argued that the missing learning mechanism was essentially backpropagation in a simulated neural network. If a connectionist system, equipped only with general learning capabilities, can learn language, then Chomsky would be wrong. That would not make Skinner right, however.

The Rumelhart and McClelland connectionist model of language learning challenged the Chomsky position as well as the approach suggesting that language learning is represented by explicit rules. They argued that they had, indeed, found a learning mechanism that could learn "impossible to learn" features of language. Because of resource constraints, and other limitations, their project was just a small piece of what they hoped to eventually generalize to other language problems like parasitic nouns. It was a kind of down payment on a full language learner, or so they hoped.

Rumelhart and McClelland trained a multilayer perceptron to learn how to form the past tense of verbs. Given an input word, such as "like," it would learn to produce the past-tense "liked." Given "swim," the network would produce "swam." And it would do this learning without any previous knowledge of language or any special language structures. Raw experience and a general learning mechanism would be enough. There would be no need for explicit rules or specialized brain modules.

Most English verbs form their past tense by adding "ed" or a variant to the end of the verb. Some of the more commonly used verbs, however, form their past tense in other, "irregular," ways, such as "think" changing to "thought." At an early age, children correctly produce the past-tense forms of both regular and irregular verbs. Soon after that, though, they tend to "over-regularize" some verbs, for example, producing "swimmed." Eventually, they relearn the adult forms. Rumelhart and McClelland's connectionist models showed the same pattern, first producing some correct forms, then over-regularizing and eventually settling on the correct form for regular and irregular verbs. Their argument was that starting from essentially nothing, the computer was able to learn a complex transformation and learn it in essentially the same way as children do. Therefore, they argued, simple learning mechanisms are sufficient.

Joel Lachter and Thomas Bever (1988) responded to the connectionist claims by pointing out that the connectionist models were not actually learning linguistic rules by themselves but succeeded because their designers included specific kinds of representations that implicitly "encoded" linguistic knowledge. Lachter and Bever called these implicit encodings *TRICS, The Representations It Crucially Supposes*. They argued that the only reason that the connectionist models succeeded is because of the crucial way that Rumelhart and McClelland chose to represent the language to the network inputs.

Connectionist models, like all models, necessarily include assumptions about what is to be learned and how the information presented to it is to be represented. Every model embodies certain assumptions, but they may not be articulated or even noticed. We discussed some of these assumptions in the context of the hobbits and orcs problem in chapter 7. Even learning to classify images from raw pixels relies on certain assumptions. In contrast to some claims by deep learning enthusiasts, this learning is far from theory neutral.

Given the large number of parameters that a connectionist model has to set, and the number of seemingly arbitrary decisions that go into building one of them, it is not difficult to see how it could mimic any rule pattern you like. The universal approximation theorem (Cybenko, 1989) shows that simple neural networks with one hidden layer can approximate any continuous function and represent a wide variety of phenomena when it has the right set of parameters. But the specific kinds of representations used for the training set also bias the patterns.

The effect that representations have on learning is not unique to language modeling. It is a central factor in all kinds of machine learning. The representations chosen by a network's designers critically impact what the network can do. Representations are never neutral. Designers may not be aware of the consequences of their representational decisions, but there are always consequences. The choices of representation may not, as Lachter and Bever claimed, be the sole explanation of a phenomenon, but they are typically central to what is learned.

Some of the evidence that Rumelhart and McClelland point to in support of their contention that their network learned in the same way as children do was that the pattern of errors reflected the same kind of U-shaped function in which some irregular forms are used correctly early in training, then come to be used incorrectly, and finally are used correctly again. The order in which the training examples are presented can affect the order in which these usage patterns emerge. The connectionist designer may intend to mimic the order in which children would be exposed to similar examples, but still there is a lot of arbitrariness to the selection of examples and their order.

Lachter and Bever argue that Rumelhart and McClelland's connectionist model was not actually a demonstration of the ability of a general learning mechanism to learn a language property. Rather, it succeeded because they happened to represent the problem in a way that made the learning easy. The right representation can make learning some problems easier, and this is an example of that, they said. Rumelhart and McClelland happened to choose a representation of the speech sounds that reflected just what the system was supposed to learn, and they happened to stage their training in a way that made the stages of confusion and production obvious.

Without these TRICS, the system would not have succeeded, Lachter and Bever claimed.

Children learning to produce past-tense verbs are initially too young to read, so it would not have made any sense to represent the words using English letters. Instead, Rumelhart and McClelland tried to represent the speech sounds of English, the phonemes, for the network (see chapter 6). The input to the network was one set of speech sounds, representing the present tense of some verbs, and the output of the network was supposed to be a different set of speech sounds, representing the past tense of these verbs.

Linguists describe the transformational rules relating present-tense and past-tense verbs in terms of phonological rules. Phonological rules describe how the sound forms of words are related to one another. The phonological rules describe the changes in speech sounds; they do not necessarily describe the spelling of those words. For instance, if a present-tense verb ends in "d" or "t," then the rule says to add "ed" to form the past tense. "Mat" is changed to "matted," and "need" is changed to "needed." If the verb ends with certain sounds, such as "sh" or "k," then just add the sound "t"—for example, "pusht," "kickt" (pushed, kicked). If the present-tense form ends with a vowel or with "b," "g," "j," or "z," then just add "d" to its pronunciation, for example, "buggd," "skid" (past tense for "ski"). Lachter and Bever review other rules as well, but these are sufficient for our purposes.

Rumelhart and McClelland chose to represent their inputs and outputs using a representation called a "Wickelphone." A Wickelphone is a representation originally described by Wayne Wickelgren, in which each letter sound is represented as a triple sequence of symbols. The first symbol in the triple represents the sound immediately before the target sound, the second represents the target sound, and the third represents the sound immediately after the target sound. The symbol # indicates the space between words. So the word "bet" could be represented phonologically as #Be + bEt + eT#. According to Lachter and Bever, the Wickelphone implicitly embeds much of the phonology of English because of this contextual structure.

This triple pattern ensures that there can be only one order for most sequences of Wickelphones. The word in the earlier example must start with B because the first triple includes the start-of-word symbol, it must end with T because the triple includes the end of word symbol, and it must have E in the center. Each input to the network was further analyzed to represent each Wickelphone as a set of Wickelfeatures, which indicated other characteristics of the word and its pronunciation—for example, whether the sound is interrupted or not, whether it is a vowel. These are precisely, said Lachter and Bever, the features that figure in the phonological rules of human past-tense formation.

Children, of course, do not start off knowing these rules; they have to learn them from the speech that they hear. They don't know anything about vowels or interruption. They do not even know with any confidence where the separations are between words. What adults hear as pauses between words can actually be briefer than some of the pauses within a pronounced word, but the Wickelphone flags where words begin and end, another advantage not available to children.

These assumptions and others, having to do with how examples are chosen and the order in which they are presented, were very influential in determining the outcome of Rumelhart and McClelland's training. Like many machine learning situations, the representation of the problem is a critical part of determining what the machine will learn. Representations like Wickelfeatures are, in Lachter and Bever's view, the representations their network crucially supposed.

Although Lachter and Bever intended their analysis to be a critique against the appropriateness of general learning mechanisms, I don't think that they succeeded in ruling out the potential for general mechanisms to learn these linguistic properties. At best, they succeeded in showing the Rumelhart and McClelland's model was not compelling evidence that a general learning mechanism is sufficient to learn language.

Lachter and Bever are not alone in suggesting that intelligence may require some special mechanism. If the linguists are right, that certain tasks require talents beyond what learning mechanisms can provide, then that calls into question the possibility of achieving artificial general intelligence. On the other hand, if general learning mechanism are sufficient, then the computer scientist need only find the right experience and represent it in the right way.

If general intelligence depends on talent or some other special capabilities, then our models have to somehow include that talent. What talent would even mean in the context of computational intelligence is itself a challenging question. Is it a predisposition? Is it a set of biases? Is it knowledge that exists prior to experience? Would talent be related somehow to the structure of the computational system?

The need for talent would not make computational intelligence impossible, but it would add the complexity of implementing talent to the list of things we must accomplish. Are there properties of the human brain that render it specifically capable of solving certain problems? What are those properties? Where do they come from? What would it take to emulate them? These are challenging problems that have not yet received any serious attention from computer or cognitive scientists.

There are always naysayers that argue that computers will never be able to do something for some reason. Many people seem to have the need to claim some special status for humans. The philosopher John Searle, mentioned earlier in the context of the Chinese room thought experiment, claims that only brains can have minds, because computer programs are purely syntactic. They cannot represent the meaning of the symbols. He is not clear, however, about just what property of brains allows them, and not computers, to have symbols that stand for, that represent objects in the world.

The philosopher Hubert Dreyfus has argued that the symbolic approach to AI, championed originally by Newell and Simon, could not work because of its inability to capture enough commonsense knowledge in a computer and because of the "frame problem"—that the world can change and the computer would have no way to identify which parts of its representation would need to be updated.

The symbolic approach to AI, Dreyfus said, was simply a recapitulation of the Western philosophical tradition from Descartes onward that held that the mind had some fundamental atomic units of knowledge and that concepts were rules and so forth. In Dreyfus's view, a computer, presumably a robot, detects the state of the world at a certain point in time to form what he calls a representation. I think that he means a symbolic description of the world at that point in time. As the computer works, it operates on that symbolic description with no further contact with the world. At some point, I think, the computer does try to sense the world, finds out that it is different from the condition that it symbolically predicted, and cannot figure out how it is different. If that is what he means, it seems like a caricature of a Jetson robot, not like any machine that would actually be used.

Twentieth-century philosophers, for example, Heidegger and Wittgenstein, recognized that that the atomic symbolic approach was bankrupt, but symbolic AI was just mimicking this bankrupt approach, Dreyfus says. That is why, he claims, symbolic AI must fail.

Connectionist approaches to AI fail for another reason, according to Dreyfus. Connectionist approaches fail because they cannot actually learn, he thought, and because even having infinite processing power will not allow them to achieve what the brain does. He argues that AI is still trying just to mimic intelligent behavior, but it cannot succeed at being truly intelligent unless it gets beyond the level of the behavior of artificial systems and gets to the level of doing what the mind actually does. He ridicules the idea that once computers have enough bits and CPUs that they can emulate the brain, they will somehow achieve consciousness and true intelligence. Computational capacity is not the answer. No one knows what the answer is. The hardest problem is how matter could ever produce consciousness, and AI and the use of computers is not helping understand it one bit, he says.

Dreyfus's view of artificial intelligence seems to start with Newell and Simon and their physical symbol systems and end with Rodney Brooks, who founded iRobot, the maker of the Roomba vacuum (among other robots). Brooks argued that robots would not need any kind of internal representation, but, according to Dreyfus, at least, he also argued that insects do not learn and automatic vacuums did not need to learn either. Dreyfus took the absence of learning in these two systems as evidence that learning was an intractable problem for AI. Somehow, he missed the entire fields of machine learning and connectionist systems that learn very well.

Dreyfus argued that knowledge is not stored as symbols and rules but as readinesses to act. Human minds do not so much store facts about the world, as in commonsense knowledge, but are instead modified to be ready to act. Knowledge is "knowing how," rather than "knowing that," as Gilbert Ryle would have put it.

Dreyfus argued that the way people learn is that the world "looks different to us" as a result of our experience. Without consciousness, there can be no meaning to "looks different to," because it has to look different to something or someone. That is, there has to be some consciousness there to be the object of "looking different to me."

It is not clear to me just what causal role consciousness might play in intelligence, even for Dreyfus. In Dreyfus's view, it somehow plays a role for the human in identifying the context in which he finds himself. In computer terms, it lets him pick the contextual frame to understand what is relevant and what is not.

Many philosophers find the question of consciousness to be the central question of philosophy. How do we get from mindless neurons to a mind that not only functions but is aware that it is functioning? I don't just respond to events in my environment—I *experience* them. There is something that it is like to be me, but only I am directly aware of just what that is. By extension, I expect that there is something that it is like to be you as well. I assume that you have experiences just as I have experiences. But in your case, I have to infer those states by analogy because I and only I have access to my own experience.

There are many theories as to exactly what consciousness is and what it does, but I don't know of any that attribute a role for consciousness in intelligence. Dreyfus hints at a role, but he is just not at all clear about what that role is. At best, he posits something akin to Descartes's famous claim, "I think, therefore, I am." If there is thinking, there has to be something that thinks, but it is not clear that that something has to be conscious.

Just as there is something that it is like to be me, something that it is like to be you, and something else that it is like to be a bat, so, too, there might be something that it is like to be a computer. From the point of view of a human, it might be boring to be a computer or mysterious to be a bat, but there could still be something. But again, there does not appear to be any causal role for consciousness in intelligence. Consciousness seems to be a red herring for artificial intelligence, and probably for philosophy as well.

Dreyfus was right about the inadequacy of physical symbol systems to serve as the basis for AI. He appears to have been right about the difficulty of capturing a finite body of knowledge for commonsense reasoning. If he is right, though, it appears to be completely for the wrong reasons. He is certainly wrong about computers not being able to learn, unless he means something idiosyncratic by the word "learn." The jury is still out on whether the relevance problem that he raises is even pertinent to computational intelligence. Most AI programs address relatively limited small worlds, and it is an object of faith that the same processes can be extended to more flexible environments.

As Lachter and Bever pointed out, the system designer is the one who selects what features of the environment are important to solve the problem at hand. As intelligence becomes more general, there may be more need for the computer to be able to select relevant data, but we have not yet reached that point.

Selecting relevant information is also an unsolved problem for humans. As discussed in chapter 7, any two objects are similar on an infinite number of dimensions and different on an infinite number. How people select the basis for comparison is itself something of a mystery. The problem is not unique to machines.

I wonder what Dreyfus would have said about self-driving cars. They represent many different kinds of contexts. Their ability to navigate through desert passes or crowded urban side streets would imply that they can, in fact, learn and can switch contexts when needed. The multiple sensor systems allow them to use different kinds of information in different situations, for example, at different speeds. They operate on a mixture of direct sensation of the world and internal representations.

Finally, Dreyfus is wrong about the necessity for consciousness. There are many reasons that could cause him to be wrong about consciousness. One of the simplest is that the way that we think (presumably consciously) we solve problems is very often not the way that we actually do solve them. We thought that chess playing would require deep knowledge about strategy and tactics. Instead, we found that a specific kind of tree search algorithm would beat the best human players. We thought that playing go could not even be practically computed because the same kind of graph algorithm used for chess simply had too many branches. A simpler algorithm again turned out to be all that was necessary. We thought that catching a batted baseball would require computing a parabolic trajectory when really all it requires is positioning one's head to keep a ball in constant apparent motion.

Daniel Dennett, another philosopher, brings up another problem with consciousness that we might call the mysterious case of Marilyn Monroe wallpaper. If you go into a room where the wall is covered with repeated images, for example of Marilyn Monroe, perhaps like those painted by Andy Warhol, you are conscious of seeing all of those repetitions. But the evidence is clear that you do not and, in fact, cannot see them all. You come to that conclusion in less time than your eyes could possibly scan the whole wall. If some of those images are changed while your eye is moving in a saccade, one of the jumps that occurs frequently as your eye moves, you are extremely unlikely to be able to notice it.

Another bit of evidence that consciousness is not the main controller of what we see is called change blindness. When someone looks at a picture, say of a large airplane, they report that they see the whole picture. They claim to be conscious of seeing the whole picture, yet it is easy to show that even large parts of it are not available. Very large changes often go unnoticed. Consciousness, it appears, in contrast to Dreyfus's claim, is itself not very good at selecting those representational features that need to change when the world changes.

Change blindness can be easily demonstrated by showing two pictures in alternation, separated by a brief flash. The flash is used to prevent the use of apparent motion cues between the two pictures. Apparent motion is the perceptual phenomenon that allows a series of still pictures presented in rapid succession to be seen as a continuous movie. The eye and brain have specific detectors for detecting motion.

In change blindness, the pictures are identical, except for one selected difference between them. One of my favorites shows a large plane. In one version, an engine is present near the middle of the scene. In the other, the engine is absent. People often have a difficult time identifying what the difference is between the two alternating images. Once you see it, however, it is obvious and difficult to stop seeing it:

https://www.cse.iitk.ac.in/users/se367/10/presentation_local/Change%20Blindness.html

http://nivea.psycho.univ-paris5.fr/ECS/kayakflick.gif

Change blindness can also be seen when a distracting event, such as a mudsplash, occurs between the changes:

http://nivea.psycho.univ-paris5.fr/CBMovies/ObeliskMudsplashMovie.gif

Changes between cuts in a video can also produce change blindness:

https://www.youtube.com/watch?v=ubNF9QNEQLA

Finally, people do not notice a change even in who they are talking to when a distracting event occurs:

https://www.youtube.com/watch?v=vBPG_OBgTWg

The point of these change blindness findings and other experiments is that our conscious awareness does not reliably reflect the environment. Rather than being an arbiter of what is relevant in a situation, consciousness is a product of what we have otherwise "decided" is relevant. We have not seen all of the images of Marilyn Monroe, but we decide that they are there and we claim consciousness of their existence even if we are wrong.

I don't believe, therefore, that consciousness is likely to be a necessary property of computational intelligence. In any case, there are simpler ways to choose which actions to perform. In the absence of compelling evidence, and philosophical intuition is not evidence, we should prefer the simpler mechanism over the more complicated one. The claim that consciousness is necessary to computational intelligence could be correct, but extraordinary evidence should be required before we jump to that conclusion.

To the extent that consciousness is relevant, my view is that it is an observational or perceptual process of our brain's functions. We do not have access to most of the direct operation of neurons, but we do have access to the eventual product of some of these operations. The reason that I have special access to my processes is simply because they are in my head and not in yours. The reason that I experience them, in other words, is because they are in me. My digestion deals with food that I ate, not food that someone else ate. My perception of my thought processes deals with action of my brain, not someone else's.

If I have too little of the neurotransmitter serotonin in my brain, I feel depressed. The amount of available serotonin changes my brain activity, and that changes my conscious feelings. Hallucinogens change brain activity, and the effect that these hallucinogens have on what we call consciousness can be profound. The idea that consciousness is some mysterious thing that needs explaining is a legacy of dualism, that mind and body are separate categories of things, as described, for example, by Descartes. There need be nothing mystical about consciousness. Dreyfus, in other words, is falling into the same Western philosophy trap that he argued was the cause of failure for symbolic approaches to AI.

Consciousness is just a natural process. If it needs explaining at all, the part to explain is why we can describe some of the brain's operations and not others. This question is, of course, complicated by the fact that our descriptions do not always match what is actually happening either in the world or in the part of it enclosed in our own skull.

Consciousness is something of a reconstruction of some aspects of our brains' functioning. But much of our intelligence may come from unconscious processes. These are processes that we find difficult to describe. According to some, consciousness is the story we tell ourselves to make sense of the world. It is a rational reconstruction of what we think has been going on in our heads, not the cause of what is going on.

Dreyfus's criticisms are largely directed at the notion of a physical symbol system hypothesis (see chapter 3). The idea is that a system that contained and operated on symbolic representations of knowledge was both necessary and sufficient to implement intelligence.

Physical symbol processing was ideally suited to the complex intellectual problems that could be described in terms of rules that operated on symbols. The belief was that it would be possible to describe every aspect of intelligence in sufficient detail that it could be implemented by a machine. This belief turned out, I think, to be misleading and inaccurate. Characterizing intelligence as symbol and rule processing and focusing on tasks with explicit rules mutually reinforced one another. By picking example problems that were easy to assess, computational intelligence designers also picked example problems that were suitable to the tools that they had developed. Their success on these problems led them to believe that these were the only kinds of problems that needed to be solved, so the tools appropriate to solving them were the tools needed to solve any other problem.

Although the AI tools and approaches have evolved significantly since Newell and Simon, this general framework has persisted. Researchers focus on problems that they can address in a reasonable amount of time. They work on problems of limited scope. They have to publish papers, get degrees, and so on. They have to bring products to market. They then overgeneralize the solution that worked to address these limited problems and think that the same approach can extend to other, unaddressed problems. It may take more processors than are currently available or more memory, or more examples, but the methods that work on limited-scope problems should be extensible other situations. Rumelhart and McClelland assumed that the approach that they used for past-tense learning would apply to other linguistic problems as well. But, if their solution depended on special-purpose representations, such generalization would not be forthcoming. A computer program that learns the past tense using special representations would not be of much use to solve other problems that require different representations.

Representations for General Intelligence

The gist of the argument is that learning under the current framework does not require special-purpose learning mechanisms but special-purpose representations. Lachter and Bever were right; these are the representations that a problem critically supposes. These representations depend on problem-specific knowledge, but general intelligence requires general representations, which we have yet to figure out.

Another problem with the physical symbol system approach is the premise that we can describe the world with sufficient specificity. That idea presumes that there are a finite number of objects in the world and a finite set of rules to deal with those objects. This assumption works when we limit the environment to certain small constrained worlds, such as theorem proving, block stacking, or games. But, when we get to less constrained worlds, it rapidly falls apart. Categorization is complex and, as we discussed earlier, difficult to represent by similarity. Like games and furniture, many categories, such as rich, smart, or tall, do not have fixed definitions.

Only 4% of people at the top of the US economy (those worth a million dollars or more) consider themselves to be rich. The other 96% of millionaires describe themselves as middle class or upper middle class. Only 11% of people with a net worth of $5 million or more consider themselves to be rich.

How rich, smart, or tall do you have to be to be labeled rich, smart, or tall? Generally speaking, it depends. Rich people are those worth more than I am. Tall people are those who are taller than I am, and smart people are those who are smarter than I am.

As discussed in chapter 7, there are ad hoc categories (such as things to take on vacation or things that make an appropriate Mother's Day gift). These ad hoc categories have many of the same properties as more common and well-used ones (for example, they each have prototypical examples). Because they are ad hoc, they cannot have been stored as a prestructured representation. Rather, the representation has to be constructed at the time, and it may not persist after its single use.

Other categories, named with two words, can also be complex and difficult to represent consistently. Consider a gummy worm, an earthworm, and a wax worm. You might like to eat the first; you would probably not want to eat the other two. An earthworm lives in the earth, and a wax worm, unlike a wax apple, is a worm (actually a caterpillar larva of the wax moth) that eats beeswax, and, as recently discovered, polyethylene plastic.

If we constrained ourselves to well-chosen examples, computers could probably manage these strange categories, but there do not seem to be a finite number of them. Birds fly, but penguins and ostriches do not. Whatever representations we come up with, they will have to be able to deal with strange categories and exceptions.

Symbol systems do not scale well. If there can be millions or more categories and each category can involve many thousands or millions of rules, it is not clear where the effort would come from to generate those categories and rules.

Inherent in this discussion of consciousness, intelligence, and the structure of categories is the critical role that representation plays in intelligence. Political arguments can be won or lost based on exactly how they are represented. Was the invasion of Iraq in the 1990s like the invasion by US forces of Vietnam, or was it like World War II? As discussed earlier, Kahneman and Tversky found that describing the outcome of a choice relative to a win (the number of people who survived because of a treatment) led to higher preference for the treatment than describing it relative to a loss (how many people would die despite the treatment) even when the numbers were actually identical in both situations.

The mutilated checkerboard problem is easy to solve with one representation, as a parity problem, and difficult to solve with another, as a layout problem. The game of go was computationally intractable when considered as a general tree problem but was more tractable when considered as a pattern-recognition problem. As in human cognition, how a machine learning problem is represented is critical to how or even whether it can be solved.

There have been some attempts, particularly involving deep neural networks, to come up with generic representations and systems that can learn their own representations. I will argue that most of this work is simply mistaken.

For example, Yoshua Bengio, Aaron Courville, and Pascal Vincent correctly note that the success of machine learning algorithms depends strongly on the representation of the data. Specific representations can make some distinctions easier to recognize and others more difficult. They also correctly recognize that selecting or constructing these representations can require a great deal of effort. It would be useful if machine learning could be used to reduce the needed effort, if it could be used to derive its own appropriate representations for solving problems. Such a capability would eventually be necessary for a truly general computational intelligence.

Consider some machine learning problem, such as categorization—for example, to categorize documents by topic, pictures by whether they contain cats, or loan applications by whether the applicant is creditworthy. Each item to be categorized could be represented by a bundle of measurable characteristics or features. The features for a document categorization problem could be the words in the documents, or sequences of letters, or some combinations of these. The features of pictures could include the pixels in the picture, the presence of some specific shapes, or something else (for example, the parameters of a discrete cosine transform, a method for summarizing groups of pixels). In evaluating loan applications, the features could include any of the characteristics we know about the applicant. Some of these features will be relevant, and some might not be.

Whatever object we are classifying, we usually put the features and objects into a table, where the rows are objects and the columns are the features. If, for example, a document has the word "teacher" in it, then the row for that document will have a 1 in the column for the word "teacher." Words that appear in the document will have nonzero values for that document's row, and words that do not appear in that document will have a zero value in the column corresponding to that word and the row corresponding to that document. The more features there are, the more combinations of features need to be considered when optimizing, and the number of combinations grows much faster than the number of elements to combine—the curse of dimensionality.

Many learning algorithms benefit from reducing the number of features that are used, provided that the preserved features are informative about the distinctions that are to be learned. There are statistical methods that are widely used to distinguish the relevant features on the basis of the information that they convey. These methods can select among the features in the initial representation, keeping only those that provide information about the category.

For example, a document that contains the word "lawyer" is also likely to have words like "judge," "attorney," or "court." If we are interested in distinguishing documents about legal matters from others, it may not matter which of these words appear in it, so long as one or more of them does. Each of these words, we might say, contributes to the topic or dimension of "legal matters." These methods can be generally described as organizing features into combinations that are most informative about the categories being trained.

Several statistical techniques can capture the correlation among features and reduce a large number of basic features to a smaller set of features. In some representations of documents, for example, the words that make up the original features may be combined into topics. Hundreds of thousands of words may be statistically reduced to a smaller set of perhaps hundreds of topics. It is usually much easier to learn about a few hundred features per object than about several hundred thousand, and these techniques ensure that these reduced dimensions reflect most of the information that could be conveyed by the full set of features.

Some of the deep learning projects claim that they can solve the curse of dimensionality and solve the need to construct representations by having the network learn representations. For example, they may use probabilistic models, autoencoders, restricted Boltzmann machines, and a few other techniques to learn the important features. The details of these techniques are not critical to the discussion at this point. These techniques do not, as I see it, actually learn new representations; they perform some of the same familiar statistical techniques that have been used in the past to transform the inputs. They combine, select, and summarize; they do not create new representations. They cannot, on their own, decide to use Wickelphones to represent words.

The statistical technique that they perform is determined by the particular deep learning structure that is designed into the network. For example, an autoencoder is a network process that takes a "raw" input, such as the pixels in a picture, and is trained to reproduce those pixels as an output, while passing the patterns through a much smaller intermediate layer.

A 200 × 200 pixel image would have 40,000 inputs and 40,000 outputs for the autoencoder. A single-layer network could accomplish this reproduction perfectly by simply passing the input through to the output, but nothing would be gained. Instead, the autoencoder includes a layer that has substantially fewer units, so, in order to reproduce the input effectively, it has to find the relationships in the input that can be summarized by these "bottleneck" units and still reproduce the outputs. In other words, this hidden layer finds the same kind of statistical summary like that used in other forms of dimension reduction, such as the topic transformation described earlier. Mathematically, this hidden layer does what is called a "principal component analysis." The patterns learned by this hidden layer are not arbitrary. They conform specifically to well-known statistics.

There are many methods that can be used to produce a bottleneck. Using an autoencoder is a choice made by the designers of the network, and the choice of the autoencoder determines the kind of representation that the network will learn. Alternatively, they could have chosen a restricted Boltzmann machine, which learns a different statistical summary, a factor analysis (see chapter 2).

In short, it is the designer of the network who decides how the inputs are to be represented. The network may choose the specific values that the hidden units will take on dependent on the data, but this is exactly the same as when a separate statistical analysis is performed on the data using any other technique. Having the network conduct this statistical analysis, rather than using one of the other well-known methods, may be inefficient in terms of the number of iterations the system must go through, that is, the number of times it must consider each input. The computational complexity may be higher than with other methods, but all of these methods have the interesting property that they do not require labeled examples. Their error can be measured without any human labels by evaluating the quality of the system's reproduction of the input.

Whether the transformation of the input is done before it is submitted to the network or whether that capability is added to the network itself does not change the fact that the transformation is an essential part of what is learned. There is no magic to having a network perform the transformation. We may say that the network learns the transformation, but actually, what it learns is the value of the transformation, not the process. The transformation is not optional, and its type is not optional to be selected by the network.

Conclusion

The representation chosen can have profound implications for the success of a machine learning project. Of the factors that affect machine learning, representation of the problem and of the input data is arguably the most important. At this point, the choice of representation is still determined by the designer of the machine learning system, but a truly general computational intelligence will require the ability to create its own representations. We are very far from achieving that in an automated way.

Resources

Bombelli, P., Howe, C. J., & Bertocchini, F. (2017). Polyethylene bio-degradation by caterpillars of the wax moth *Galleria mellonella. Current Biology, 27,* R292–R293. http://www.cell.com/current-biology/fulltext/S0960-9822(17)30231-2

Chomsky, N. (1959). A review of B. F. Skinner's *Verbal Behavior. Language, 35,* 26–58.

Chomsky, N. (1965). *Aspects of the theory of syntax.* Cambridge, MA: MIT Press.

Chomsky N. (1975). *Reflections on language.* New York, NY: Pantheon.

Chomsky, N., & Gliedman J. (1983, November). Things no amount of learning can teach: Noam Chomsky interviewed by John Gliedman. *Omni, 6*(11). https://chomsky.info/198311__/

Cybenko, G. (1989). Approximations by superpositions of sigmoidal functions. *Mathematics of Control, Signals, and Systems, 2,* 303–314.

Dipshan, R. (2017). Why artificial intelligence can't compete with humans, and vice versa. LegalTech News. http://www.legaltechnews.com/id=1202783879709/Why-Artificial-Intelligence-Cant-Compete-With-Humans-and-Vice-Versa

Dreyfus, H. L. (2007). Why Heideggerian AI failed and how fixing it would require making it more Heideggerian. *Philosophical Psychology, 20,* 247–268. http://leidlmair.at/doc/WhyHeideggerianAIFailed.pdf

Dreyfus, H. L. (2013). Why is consciousness baffling? https://www.youtube.com/watch?v=Bhz7bRiuDk0

Dreyfus, H. L., & Kuhn, R. L. (2013). Artificial intelligence–Hubert Dreyfus–Heidegger–Deep Learning. https://www.youtube.com/watch?v=oUcKXJTUGIE

Dubey, R., Agrawal, P., Pathak, D., Griffiths, T., & Efros, A. (2018). Investigating human priors for playing video games. Proceedings of the 35th International Conference on Machine Learning, in *Proceedings of Machine Learning Research, 80,* 1349–1357.

Evans, V. (2014). Real talk: The evidence is in, there is no language instinct. Aeon. https://aeon.co/essays/the-evidence-is-in-there-is-no-language-instinct

Frank, R. (2015). Most millionaires say they're middle class. CNBC. http://www.cnbc.com/2015/05/06/naires-say-theyre-middle-class.html

Lachter, J., & Bever, T. (1988). The relation between linguistic structure and associative theories of language learning—A constructive critique of some connectionist learning models. *Cognition, 28,* 195–247. doi:10.1016/0010-0277(88)90033-9; https://www.researchgate.net/publication/19806078_The_relation_between_linguistic_structure_and_associative_theories_of_language_learning-A_constructive_critique_of_some_connectionist_learning_models

Nagel, T. (1974). What is it like to be a bat? *The Philosophical Review, 83,* 435–450. doi:10.2307/2183914 JSTOR 2183914; http://www.jstor.org/stable/2183914

Pepperberg, I. M. (1999). *In search of King Solomon's ring: Studies to determine the communicative and cognitive capacities of grey parrots.* Cambridge, MA: Harvard University Press.

Plunkett, K., & Juola, P. (1999). A connectionist model of English past tense and plural morphology. *Cognitive Science, 23,* 463–490. http://onlinelibrary.wiley.com/doi/10.1207/s15516709cog2304_4/pdf

Rensink, R. A., O'Regan, J. K., & Clark, J. (1997). To see or not to see: The need for attention to perceive changes in scenes. *Psychological Science, 8,* 368–373.

Rumelhart, D., & McClelland, J. (1986). On learning the past tenses of English verbs. In D. Rumelhart, J. McClelland, & the PDP Research Group (Eds.), *Parallel Distributed Processing* (Vol. 2) (pp. 216–271). Cambridge, MA: MIT Press.

Savage-Rumbaugh, E. S. (1986). *Ape language: From conditioned response to symbol.* New York, NY: Columbia University Press.

Savage-Rumbaugh, E. S., Murphy, J., Sevcick, R. A., Brakke, K. E., Williams, S. L., & Rumbaugh D. L. (1993). Language comprehension in ape and child. *Monographs of the Society for Research in Child Development, 58,* 1–221.

Searle, J. R. (1984). *Minds, brains and science.* Cambridge, MA: Harvard University Press.

Terrace, H. S., Pettito, L. A., Sanders R. J., & Bever, T. G. (1979). Can an ape create a sentence? *Science, 206,* 891–902.

Wallman, J. (1992). *Aping language.* Cambridge, UK: Cambridge University Press.

10 Algorithms: From People to Computers

Algorithms and heuristics are important because they show what unaided human brains do and because they show what more those brains can do when they are deployed systematically. Human intelligence depends on both heuristics and the algorithms that have been invented over the last 50,000 years.

Humans have become increasingly intelligent over the last 50,000 years because they have invented, implemented, and followed procedures that make thought more systematic and effective. The use of algorithms has made human thought more effective and has made automatic computer processes possible.

The word "algorithm" comes from the Latin word *algorismus*, which is a Latinized form of the name Al-Khwarizmi, a Persian mathematician from the ninth century, and the Greek word *arithmos*, which means number. The word came into more prominent use during the thirteenth century in the context of changing over to the use of so-called Arabic or Hindu numerals (the ones we use today) from Roman numerals.

Roman numerals are well suited to dates and to counting objects, but they are extremely limiting in other ways. Just multiplying two numbers in Roman numerals is a multistep process that is prone to errors. Many kinds of mathematics were only possible when the representation of numbers changed to the positional decimal one we have today. It may seem strange to talk about algorithms for multiplying numbers, but, in fact, this is an active area of research, particularly for how to effectively multiply very large numbers.

Most of us learned the carry method in elementary school, where two multidigit numbers are placed in a column and each digit in the top number

Multiplying Roman Numerals

Multiplying two numbers like 21 × 17 is easy using Arabic numerals but a very complex task with Roman numerals. To multiply XXI × XVII:

Make two columns, and write XXI in one column and XVII in the other.

Divide the number in the left-hand column by 2, *ignoring the remainder* (XXI → X).

Multiply the number in the right-hand column by 2 (XVII → XXXIV).

Repeat the above two steps until the left-hand column contains I.

X → V; XXXIV → 2 = LXVIII

V → II; LXVIII → CXXXVI

II → I; CXXXVI → CCLXXII).

Go down the rows and strike out each line where the number in the left-hand column is even.

XXI	XVII
~~X~~	~~XXXIV~~
V	LXVIII
~~II~~	~~CXXXVI~~
I	CCLXXII

Add the remaining values in the right-hand column (XVII + LXVIII + CCLXXII = CCLLXXXXVVIIIIII = CCCXXXXXVII = CCCLVII = 357).

No comparable method exists for dividing Roman numerals.

is multiplied by each digit in the lower number and the result is added up. Multiplying two 3-digit numbers takes 9 single-digit multiplications, and multiplying two 4-digit numbers takes 16 multiplications. Multiplying two 10,000-digit numbers takes 100 million single-digit multiplications.

In 1960, Anatoly Karatsuba developed a method that combined multiplications with additions and subtractions to greatly reduce the number of single-digit multiplications that would need to be done. Karatsuba's method would multiply those 10,000-digit numbers in a little more than 2 million operations.

In 1971, Arnold Schönhage and Volker Strassen found an even faster method that would reduce the number of operations needed to multiply two 10,000-digit numbers to 204,500. A still faster algorithm was discovered by David Harvey and Joris van der Hoeven in 2019 that reduced the number of steps to about 92,000 operations. The specific algorithm can make a very big difference in how long it takes to do basic computations, such as finding the next prime number. The difference is so large, in fact, that it can mean the difference between accomplishing some feat and not being able to live long enough to see it through to completion.

Not all algorithms are necessarily numeric. Recipes can be thought of as algorithms. If you follow the steps given in the recipe, the result will be the dish that you expect.

Although humans are capable of brilliant episodes of genius, people spend much of their time at far lower levels of intellectual challenge. Herbert Simon argued that people generally *satisfice* rather than optimize what they do. He thought that a fully rational approach would leave people lost in thought contemplating unreachable alternatives. Instead, Simon argued that they consider the choices that are readily available and let that be good enough.

Yet, when we need or want to be intelligent, we seem generally capable of achieving it. We are capable of being thoughtful. What I think that means is that we are capable of engaging in systematic analysis, even if we do not always do so. Humans have developed thinking tools, power tools for the mind, if you will, that let us achieve (occasional) soaring levels of genius. These tools help us to plan strategies and make better use of the information that we have. Similar tools are used in computational intelligence where they are also highly effective. Think of these as tools for highly effective intelligence.

The distinction between everyday thought and effective thought corresponds roughly to Daniel Kahneman's distinction between System 1 and System 2 thought. Kahneman objected to the traditional view of humans as rational decision makers—the so-called rational man view from economic theory. In this traditional approach to economics, people were thought to generally choose alternatives based on their own self-interest. When a person did not choose the best alternative, it was a rare occasion on which emotions intervened.

This rational view of decision making is not particularly good at explaining human behavior. Remember, for example, that more people will take advantage of the lower price when it is characterized as a late payment penalty than when it is described as an early payment discount. People would rather buy hamburger meat that is 90% lean than hamburger meat described as 10% fat.

In the rational-man view, departures from rationality are viewed as errors. Instead, I have argued that the nonrational decisions are principled in their own way. More importantly, I think that they are not errors at all, but indicators of a kind of thought that is derived from natural intelligence, and essential machinery for the creation of human intelligence.

System 1, in Kahneman's view, is fast, based strongly on recognition, automatic, and emotional. System 2 is deliberative, logical, effortful, and systematic.

System 1 is used for such tasks as:

- recognizing that one object is farther away than another
- recognizing the emotion being displayed in a photograph of a person
- recognizing that there are four coins on a table (without having to count them)
- solving simple arithmetic problems, such as 2 + 2

Most of our everyday activity is governed by System 1. Most of what we do on a daily basis is habitual and familiar. Given a familiar math problem like 2 + 2 = ?, the answer is directly available. The answer is the result of crystallized intelligence. If we see a person coming toward us with a smile on his face, eyes broad, mouth turned up, we do not have to exert much effort to expect that person to be happy. If we hear a strange bump in the night, we don't have to calculate to expect that something dangerous is going on. If someone says, "How are you?" our instant response is "Fine." These situations are clichéd for a reason; they recur frequently, and a memorized response is usually enough.

System 2, on the other hand, is used for such tasks as:

- solving complicated arithmetic problems like 13 × 27
- deciding whether to accept a job offer
- monitoring the appropriateness of behavior at a party
- parking in a narrow parking space
- verifying an unfamiliar logical syllogism
- evaluating complex legal arguments

System 2 processes are among those that are most closely associated with intellectual achievement and intelligence. They require effort, take time, and can be challenging. System 1 often involves jumping to conclusions. These conclusions are not random, but their pattern reveals much about the process of System 1 thought.

One of the most revealing studies of System 1 thought is an experiment that Kahneman performed with Amos Tversky. They presented the so-called Linda problem to a group of college students. The students were told about a young, single, outspoken person, Linda, who was deeply concerned, while a student, with issues of discrimination and social justice. One group of people was asked to rank a list of eight scenarios by how *similar* each scenario was to the description of Linda in her thirties. They decided that Linda is a very good fit for an active feminist, a reasonably good fit for a bookstore employee and someone who takes yoga classes, and a poor fit for being a bank teller or an insurance salesperson. Critically, they said that Linda's profile more closely *resembled* the idea of a feminist bank teller than a bank teller in general.

Another group of participants was asked to judge how *likely* each statement was. How likely was it that Linda was a bank teller versus how likely was it that she was a feminist bank teller? Rationally, the probability has to be higher that Linda is a bank teller than a feminist bank teller because only some bank tellers would be feminists and there can be no feminist bank tellers who are not also bank tellers. Feminist bank teller must be a subset of all bank tellers. However likely Linda was to be a feminist bank teller, she could not be any less likely to be a bank teller.

Nevertheless, 89% of undergraduates rated it *more likely* that Linda was a feminist bank teller than that she was a bank teller. Logically, that cannot be true. Even participants who did both tasks still rated feminist bank teller as more probable than bank teller.

Kahneman and Tversky interpreted this result as a conflict between representativeness or resemblance, on the one hand, and logic on the other. Linda could resemble a feminist bank teller more than she resembled an average bank teller, but the rules of probability dictated that she could not be more likely to be a feminist bank teller than a bank teller of any stripe.

Kahneman and Tversky argued that representativeness governed the participants' choices in both decisions. Apparently, the participants made their judgment using System 1, using pattern matching, rather than basing their decision on a logical analysis of the situation.

In another study Kahneman and Tversky asked participants to decide which of these two scenarios was more likely:

- a flood somewhere in North America next year where more than a thousand people die
- an earthquake in California next year causing a flood where more than a thousand people die

As in the Linda problem, a flood in California, as part of North America, and a flood caused by an earthquake both have to be less probable (or at least no more probable) than a flood by any cause anywhere in North America. Still participants chose the California version as more probable. Earthquakes are more closely associated with California than with other states, so the California story may be more plausible sounding, that is, it may be more representative than the idea of a flood anywhere in North America.

The strong resemblance between scenarios and the participants' expectations caused them to decide on the basis of similarity rather than logic. When the similarity factor is reduced, however, people do make the logically consistent choice. For example, if they are asked which is more probable

- John has hair
- John has blonde hair

they logically choose that it is more probable that John has hair than that he has blonde hair.

Kahneman and Tversky identified other decision methods that people use when they engage only System 1 to make decisions. The Linda and flood examples use the representativeness heuristic. Another System 1 heuristic is availability—how easy it is to think of examples. If an item is easier to think of, if it is more available, it is estimated to be higher probability than an item that is less available. We discussed this heuristic in the context of judging city size and basketball playoff success in chapter 2.

If we are evaluating a risky action, for example, we will tend to overestimate the risk if we think of times when that action led to poor outcomes and underestimate the risk if we think of times when the action led to successful outcomes. Opinions are easier to believe if we can think of examples that support that opinion than if we tend to think of counterexamples.

Anchoring is a related heuristic. The context makes some things easier to think of than others. If we asked a person how old John Wayne was when he died, we will get one estimate. If we ask whether he was at least 96 years old and then ask his age, people will guess that he was older than if we ask whether he was 35 years old or more when he died.

Nassim Taleb (quoted in the 2014 Guardian article) described another example of anchoring. He told the story of trying to manage investments for clients using a strategy that would have common small losses but would also have some rare big gains. He said that clients kept "forgetting" the principles of the strategy and complained about their frequent losses. He found, however, that if he had his clients indicate, at the start of the year, how much they would be willing to pay for the chance at the big payoff, he would then post their progress over the year relative to the amount that they were prepared to pay. If they lost less, they then saw it as profit from what they expected to lose. It was money "recovered" rather than money lost. The estimate at the beginning of the year set an anchor against which all future transactions could be judged. Instead of using the full investment as the anchor, and reporting losses relative to this anchor, he reported "gains" relative to their discounted anchor.

Framing is another factor that affects how readily people retrieve examples. If people are asked whether they would choose to have surgery for a problem when 90% of people who get that surgery survive, they are more likely to choose the surgery than if they are asked whether they would get a surgery when 10% of the people getting this surgery die. Again, these two choices are logically identical, but talking about survival makes examples of survival more readily available, and talking about death makes examples of death more readily available.

On the other side, it is difficult to base decisions on information that you don't have or on estimates of future value that might be wrong. Markets are volatile, and postponing decisions until more information is available may cancel the value of that information. The profitable opportunity may have passed, and the value of a stock may further diminish while waiting to get information. At present, we do not know how important these kinds of heuristics are for general intelligence, but because these heuristics are so widespread among humans, there is a strong chance that they could be an essential part.

These heuristics, some of which are listed in table 5 (also see chapter 7), are important for thinking about computational intelligence for two

Table 5
Some Cognitive Biases/Heuristics

Cognitive "Bias"	Limitation	Potential Benefit
Small sample biases	The tendency to be swayed by small samples without regard for how representative they might be.	Ability to reach conclusions based on small amounts of evidence.
Confirmation bias	The tendency to look for information that supports your prediction rather than information that challenges it.	Minimizes the amount of evidence that would be needed to make a prediction.
Conservation	The tendency to be slow in adjusting beliefs as new contrary information becomes available.	Resistance to irrelevant information. Observations can fail to support a prediction for many reasons other than the prediction being wrong.
Hindsight bias	The tendency to believe that past events were more predictable than they were in actuality.	Avoid "paralysis by analysis." Reinforces the idea that past evidence was successfully collected and relevant.
Illusion of control	The belief that one has more control of the events that occur than is reasonable; overestimation of one's control.	Increases the motivation to find solutions to problems.
Mere exposure effect	The tendency to believe statements that have been presented repeatedly. The tendency to like familiar things.	Recurring events tend to be valid indicators. Methods worked out for familiar situations tend to be reliable.
Overconfidence effect	The tendency to overestimate one's expertise. To express unsupported certainty in one's own decisions and estimates.	Supports the idea that unproven predictions are still useful.

Note: See https://en.wikipedia.org/wiki/List_of_cognitive_biases for more biases.

reasons. First, they seem to illustrate what unaided human brains do. People have evolved the kind of skills Kahneman associates with System 1 without formal training, and perhaps without any training at all. System 2 skills, on the other hand, seem to depend on some formal training, for example, people are specifically schooled in how to conduct thorough, systematic analyses.

Second, although we can construct situations where these thinking heuristics lead reasoning astray, they are also highly likely to play an important role in overall intelligence. If we jump to conclusions, for example, we will mostly be right without having to experience thousands of examples. We do not always have the luxury of delaying useful learning to accommodate thousands of examples.

In chapter 3, we quoted Turing's (1947/1986) comment in his report to the London Mathematical Society: ". . . if a machine is expected to be infallible, it cannot also be intelligent." The very heuristics that can be fooled by clever experiments are likely to also serve as the means by which humans readily and quickly solve everyday problems. System 1 abilities may not be sufficient to support the full intellectual achievement that we celebrate as intelligence, but they are likely to be a necessary part of human intelligence. General intelligence probably requires the algorithms of System 2 along with the heuristics of System 1. It is difficult to investigate such questions, but I believe that it is important to find out more about what these heuristics do and how they contribute to overall intelligence. These biases are essentially absent from computational approaches to intelligence, and that may be a serious mistake.

Most of the progress that we would associate with intelligence has come from the adoption of algorithmic processes of the type associated with System 2. We turn next to consideration of some of these algorithmic methods.

Optimal Choices: Using Algorithms to Guide Human Behavior

Some decisions are better than others. In fact, there is a theory of optimal decision-making that can be used to guide decision making. For example, it can be used to choose a job, a mate, a secretary, or a school. It can be used to decide whether the blip on a sonar screen indicates a whale or an enemy submarine.

Optimal does not mean perfect; it means to make the best decision one can among the available alternatives, based on the currently available information. Optimal decision theory is one of the power tools that not only help machine learning systems to adapt to solve their problems but help people to be more systematic in solving their own problems.

Optimal decisions have two components. The first is the available evidence, and the second is the method of selecting the best choice among the available choices. Optimal decision theory is an ideal, in that it can be shown that no other approach can consistently do better than an optimal decision maker.

Optimal decision theory grew out of World War II research. Part of it was focused on how a radar operator should decide whether a blip on the screen was due to an enemy plane or something else. In order to improve the quality of these decisions, psychologists and engineers set out to see if they could come up with a model of how best to make that decision. Errors could be costly. Missing an enemy plane could mean that people would die. Falsely reacting to what might look like an enemy plane could waste resources, which could also lead to people dying.

For example, during the brief 1982 Falklands War between Britain and Argentina, the British warship HMS *Brilliant* torpedoed two whales and killed a third from a helicopter. Based on the evidence available, they mistook the whales for a submarine.

The experience of the *Brilliant* is exactly the kind of problem that led to the development of optimal decision theory. The sonar provides imperfect information about a potential target. The sea bottom around the Falklands is littered with old shipwrecks, whose sonar profiles are similar to that of a submarine. Unfortunately for the whales, their sonar profile was also similar. When whales came up to breathe, a flock of seagulls would gather, which caused a blip in the radar as well, further supporting the idea that it was a submarine and not a whale.

The similarity between military targets and wildlife was an important challenge for the technology and operators of 1982. The sonar or radar operator had to decide whether each blip was a return from an object that could be safely ignored or one from a potential enemy. Because of the similarity of the signals, those decisions could not be perfect, but they could be made in an optimal way.

Optimal decision theory uses Bayes's rule. The Reverend Thomas Bayes, in the eighteenth century, came up with a simple equation that describes how to update a probability estimate. Bayes's rule tells the sonar or radar operator how to make the best decision under these circumstances.

According to Bayes's rule, the decision as to whether a blip on a radar screen is a submarine versus a whale depends on two kinds of probabilities. The first probability, called the prior probability, is the probability that a submarine is in the area. It is called the prior probability because it is estimated before we get any specific evidence from our radar or sonar system. The British Navy had already scuttled one Argentine submarine and had intercepted communications that another was assigned to attack the British fleet. So they had a reasonable expectation that they would, in fact, encounter an Argentine submarine. The second kind of probability is the probability of observing the evidence, for example, the strength and character of the blip. A specific kind of blip could result from either a whale or from a submarine, but one of the two would be more likely than the other to produce that particular kind of blip. Bayes's rule describes how to combine these two kinds of probabilities to come up with a posterior probability—the probability that it is a submarine given the prior probability and after observing the evidence.

To summarize it, the more likely submarines are to be in the area, the less evidence we need from the radar or sonar to decide that we are seeing a submarine. The less likely submarines are to be in the area, the more convincing we need, the stronger the evidence we need from the radar.

The relative cost of making an error of each kind is also used as part of the decision process. In the case of radar, there is a cost to deciding that there is an enemy submarine when there is not one (a false positive or false alarm) and a different cost to deciding that there is no enemy submarine when there really is one (a false negative or miss). But if the cost of a false positive is small (the cost of scrambling interceptors, for example) while the cost of a miss is high (the submarine attacks, killing many people), then optimal decision theory suggests how to take that imbalance into account.

When information is imperfect, then an optimal decision maker will adjust the decision process to prefer the less costly error. When the information is ambiguous, an optimal decision maker will make some errors, but it can choose which kind of errors are preferable. So, in this example, an optimal decision maker would need less convincing that the blip really is an enemy submarine, knowing that there could be Argentinean submarines in the area. The *Brilliant* operators decided that it would be preferable to blow up a few whales than to let an enemy submarine get close enough to destroy a British ship.

A self-driving car faces the same kind of decision problem. If its sensors detect what looks like an obstacle in the road, it could be disastrous to run into it, but only mildly annoying to swerve or slow down to avoid it.

An optimal decision maker combines all of its available information, about the relative likelihood of events, about the strength of the evidence, and about the costs of different kinds of errors to come up with a criterion by which to make its decisions.

In general, people tend to use heuristics to make decisions. They rarely exert the effort to make optimal decisions but typically are willing to put up with decisions that are good enough. Sometimes it is critical to do better, and in fact to do the best that one could do. Under these circumstances they don't go on gut instinct or the "seat of their pants"; rather, they engage in structured processes using optimal decision theory as their guide.

John Craven used a variation of optimal decision theory to find a missing hydrogen bomb. In January 1966, two B-52 bombers were flying off the coast of Spain. Each bomber held four H-bombs as part of a Cold War program intended to deter Soviet aggression. While joining up with air tankers for in-air refueling, one of the bombers collided with its tanker. The resulting fireball killed all four crew members on the tanker and three of the crew from the bomber. The flaming wreckage fell on the village of Palomares, Spain, along with three of the four bombs. The conventional high explosives in two of the bombs detonated, leaving 100-foot-wide craters and scattering radioactive debris all around the countryside. The third bomb landed on soft ground and was largely intact, but the fourth bomb was nowhere to be found.

Concluding that the fourth bomb had landed in the sea, the US Air Force eventually requested help from the US Navy. The navy handed the project to Craven, then head of the Navy Special Projects Office. US President Lyndon Johnson was concerned that the Soviets would find the bomb and exploit it. The navy argued that it was lost forever in the sea, but Craven had already shown how he could find stuff that was supposedly lost. So he was charged with finding the H-bomb, an object the size of a canoe in a poorly mapped area of the ocean.

A fisher from the village of Palomares said that he saw a parachute coming down at about the time of the incident and told the navy where that was. They did not believe him, however, because the position he described did not match where they thought the bomb had to be and because he did not use modern navigation devices to identify his location.

Craven used a version of optimal decision theory to decide where to look. He divided up the sea near Palomares into small squares and then enlisted a team of experts to estimate the probability that the bomb fell in each of those squares based on the probabilities of events such as whether one or both of the bomb's parachutes opened. Whether it fell straight into the water or drifted with the wind. Then he set about collecting data to optimally update those estimates.

Craven's estimates suggested that the most promising places to look were far from where more conventional search techniques had predicted the bomb to be. When the bomb did not turn up in the most likely places, the team adjusted their estimates to take this evidence into account. Eventually, the navy decided to listen to the fisher's report regarding the parachute. It was a spot that Craven's team predicted to be highly likely but had not yet searched. When the deep submersible *Alvin* was sent to look for the bomb in 2,550 feet of water, it eventually found the parachute, and under it was the bomb. The fisher's report was a powerful piece of evidence. When combined with the prior probabilities estimated before his evidence, the new estimate showed a very high likelihood that the bomb would be found in a particular spot, and there it was.

Part of the innovation in Craven's search method was the recognition that searching an area might not actually find the bomb, even if it was there. For example, some of the navy equipment could only search to a depth of 200 feet, but the bottom was over 2,000 feet down. Even if this detector was right over the bomb, it would not detect it. Searching a location with inadequate equipment actually provided no evidence at all about whether the bomb was there, so it was unreasonable to remove that square from the bomb's possible location. As it turned out, the navy had been spending a lot of time searching in ways that would never be able to find the bomb even if they looked in the right location. Craven's team took the efficacy of the search into account in adjusting their probabilities.

Optimal decision theory does not always lead to the correct answer, but over the long haul, it does more often than other methods do. As a site was adequately searched, that is, searched in a way that could find the bomb if it was there, that search reduced the probability that the bomb was in that location and simultaneously increased the probability that it was in other sites that had not yet been adequately searched. Using these mathematical techniques and the suggestion of the Palomares fisher, they were eventually able to find and recover the bomb.

Two years later, Craven applied similar techniques to find the nuclear submarine USS *Scorpion*, after the sub sank near the Azores in 1968. In that same year, he again applied similar techniques to finding a sunken Soviet submarine in the Pacific.

Optimal decision theory can even be applied to dating. The goal of the so-called marriage problem is to decide when to stop dating and settle down. Stripped of its romantic implications, the very same strategy that is optimal for dating is also relevant for those making hiring decisions, for car buyers, for renters, and even for burglars.

To keep it simple, the standard version of this problem makes a few assumptions. You date potential mates in random order. Before you date a candidate (as I said, stripped of its romantic aspects), you do not have any idea how suitable she or he may be. After dating a candidate, you can reliably rank that candidate relative to all of the other candidates you have seen.

After each date, you decide whether to marry your current date or keep looking. The assumption is that you get to date only one person at a time. Once you stop dating that person, you cannot go back to an earlier candidate. You can compare the current candidate against all previous dates, but you do not know anything about dates you may have in the future. In this problem, your only decision is to date or mate. How do you know when it is time to settle down?

In optimal decision terms, this is a stopping problem. How many options should you consider before selecting your future partner? Each date has a cost (at least in terms of time if not in terms of coffee or dinner). There are two ways to fail. You can stop too soon and accept a less than perfect mate, or you can go on searching for too long, missing your one true love. These two kinds of errors correspond to the problem faced by the crew of the HMS *Brilliant*. Accept a sonar or radar blip as an indication of an enemy sub when it is not, or reject a sonar or radar blip as an indication of a whale when it is not. Each observation has a cost and each outcome has a value.

Clearly when searching for a mate, you are not likely to be interested in a date who is not the best one you've seen so far. After the first date, all you know is that your first date may be better than nothing. Your second date may be better or worse than the first, but again, you don't know much about the field until you've had some more experience. The first one may actually be your best choice, but by the time you discover that, everyone has already moved on. The third randomly selected date has a 1/3 chance of being the best yet. The fifth one has a 1/5 chance of being the best yet. So, the best date yet will become less likely, the longer you sample.

The odds of getting the best mate if you simply choose randomly will be 1 over the size of the dating pool. If you expect to date three potential partners and pick one of them randomly, then the probability of getting the best one is 1/3 = 33.3%.

One way to figure out what the optimal strategy would be is based on the idea of listing all of the different ways our experience could come out. For this problem, you don't know how the prospects are ordered, but if there are three possibilities, then you know that they must be ranked 1 (the best), followed by 2, and then 3. You just don't know which prospect goes with which rank.

There are six possible orders in which you could date the three prospects. Following the specified rules, each of these orders has a determined pick. Here are the rules:

- If your current date is better than your previous one, then choose your current date.
- If your current date is worse than your previous one, then go on to your next choice.
- If you have run out of candidates, choose your current date.

If there are three potential partners, we can list out all six of the possible scenarios. In two of the scenarios, the best choice appears first (the rank of each date is either 1, 2, 3 or 1, 3, 2). If we always choose the first candidate, therefore, our chance that the first candidate is the best one is 2/6 or 33.3%.

With three dates and three candidates, they could be dated in one of these six orders (recall that we did something similar with the three-socks problem, listing out all of the potential outcomes).

a. 1, 2, 3, mate with 3

b. 1, 3, 2, mate with 2

c. 2, 1, 3, mate with 1

d. 2, 3, 1, mate with 1

e. 3, 1, 2, mate with 1

f. 3, 2, 1, mate with 2

In scenario (a), you date all three candidates because the second one is less preferred than the first. The third is least preferred, but you have no more candidates.

In scenario (b), you date all three, but end up with the second-best candidate.

In scenarios (c), (e), and (f), you select the second date because the second one is preferable to the first one. You end up here with the most preferred candidate in scenarios (c) and (e).

In scenario (d), you date all three, but because the second candidate is less preferable than the first, you go on to the third, who turns out to be the best choice.

Following these rules, you would end up with the worst choice in one of the six scenarios. You would end up with the best candidate in three of the six scenarios, and with the candidate who would be your second choice in two of the six scenarios. In general, following the stopping rule of passing on the first date and then choosing the next date that is superior will get you the best mate in 50% of these scenarios.

That's pretty good. Just guessing will give you the best mate in a third of the scenarios and following the rules will give the best one in half of them. This strategy is still not perfect, but there does not seem to be any one that yields more success based on the information that you have.

We could similarly list out all of the 24 scenarios for four candidates and the 120 scenarios for 5 candidates and count the number of successful choices with each. But this becomes very cumbersome and error prone as the number of potential candidates increases. Instead, there is an algorithm that can be used to calculate the best strategy for any number of potential mates. In the long run, the optimal stopping rule is to use a certain number of dates just to set our standard, and then follow the rule of choosing if the next one is better than this standard and continuing with our dates if the next one is not better. The optimal number of dates to use to set the standard is about 37% of the number of candidates you think that you will date. To review, the general strategy is to date a certain number of times without making a commitment—play the field—and then commit to the next one that is better than any you have seen or give up and accept the last one.

If you follow this strategy with a pool of 30 potential choices, then you will play the field for the first 11 candidates and have a 37.86% chance of ending up with the best one in the pool, that is, the one you would pick if you had complete information about all of the candidates. With a pool of 100 candidates, you should stop after 37 dates, and you will then have a 37.1% chance of ending up with the best candidate.

Assuming that your search for a mate might extend from, say, the time you are 18 years old to the time you are 40 years old, if you date at a fairly steady rate, you can apply the same 37% approach to determine that the optimal strategy is to play the field until 26 years of age and then propose to the next candidate you meet who is better than the original pool you dated before your 26th birthday. By the way, the average age of first marriage in the United States is 28.2 years, suggesting either that young people in the United States are slightly suboptimal when intuitively choosing a mate or they estimate that their pool of potential mates is slightly larger, expecting to continue searching until they are about 46 years of age.

Optimal decision theory specifies when to stop, depending on your tolerance for uncertainty. At first, each date provides a lot of information about the dating pool, but over time, each additional date provides less and less new information. Given this strategy, you will pick the best available person with a high probability depending on the number of dates you went on when playing the field.

This problem is also called the secretary problem—deciding which secretary to hire based on interviews with a random selection of candidates. It can be proven that the 37% standard is optimal according to the algorithm, which is called the "odds algorithm."

The odds algorithm provides an optimal solution to the secretary problem, to apartment hunting, to selling your car to the best bidder, and many other kinds of situations where the problem is to determine when to stop. We often think of algorithms as being cold and heartless, but this one includes room for subjective opinion. It does not tell you what characteristics make a good mate or how much you should love a potential mate.

It is up to your subjective judgment to decide which candidates are better than others, but assuming that you have some reasonable way of deciding whether a date or anything else is interesting, in the sense of the odds algorithm, then it tells you the optimal time to stop searching and select.

Paul Meehl described an algorithm of a different sort for systematically combining subjective judgments. In 1954, Meehl published a book, *Clinical versus Statistical Prediction: A Theoretical Analysis and a Review of the Evidence*, in which he compared what he called *clinical judgment* with what he called statistical or *mechanical judgment*. Today, we might call his mechanical judgment algorithmic. He anticipated the role that artificial intelligence would play in decision making. For him, the artificial intelligence was just a rule written down as an equation, but it was still an embodiment of how it is that AI can manage to exceed the accuracy of physicians and others in diagnosis, just by being systematic.

Meehl was concerned with how psychologists use informal methods to reach their diagnoses. They would collect whatever evidence they had, including test scores, judgments about the severity of symptoms, and other information and then make a clinical judgment about the right diagnosis. For clinical judgment, read informal, intuitional, or subjective. Instead, Meehl showed that their diagnosis would be more accurate and consistent if they combined the various sources of evidence systematically, using an algorithm.

As in the marriage problem, Meehl's method did not eliminate subjective judgments; it just provided a systematic way to combine them. For example, the psychologist might have to make a subjective judgment about whether the patient's symptom was severe enough to count. Everyone has some depressed days just from ordinary experience. The psychologist would have to judge whether a patient's depressed mood was severe enough to merit a diagnosis or whether it was just an ordinary bad day. Even with this level of subjective judgment, Meehl found that combining the evidence systematically using an algorithm resulted in substantially higher diagnostic accuracy.

A clinician using informal processes might come to a different diagnosis from one patient to the next despite both patients presenting exactly the same pattern of data, but using Meehl's method, once the data were available, even a clerk or a computer should be able to come to a reliable diagnostic conclusion.

Meehl's emphasis on statistical over clinical judgment is applicable to many other forms of human judgment, including hiring decisions, courtroom assessments, and others. Meehl did not talk explicitly about optimal decision theory; it was not well-known in the 1950s. Nor did he talk about artificial intelligence. That phrase was not coined until 1956. But he did show that the kind of systematic integration done by computers can help improve the reliability of human judgment as well as its accuracy.

Modern diagnostic computer programs can outperform humans (see, for example, Esteva et al., 2017) because (at least in part) they use data in a systematic way, less affected by unconscious bias and distraction. Diagnostic computer systems are not better because they are computers; they are better because they follow specific repeatable methods. The use of these algorithms, whether executed by a computer program or by a person, can improve the quality of human performance. But these methods are also strongly data dependent.

None of these approaches, including Meehl's method, would come up with a sound diagnosis if the assessments it integrated were not themselves accurately recorded. Medical diagnostic programs, like Esteva's system for diagnosing skin cancer from photographs, would not work reliably if these systems were trained with inconsistent data. Algorithms can have an air of objectivity and authority, but their objectivity and authority depend on the data that these systems are given.

One algorithmic system based on machine learning was designed to predict the medical outcome of patients with pneumonia. Some patients with pneumonia can be safely sent home to recover; others need to be hospitalized. Richard Caruana and his colleagues built a machine learning system to try to help physicians make the decision about whom to hospitalize.

This model came up with some surprising findings. Generally, the risk of dying from pneumonia increases with age. There is a sudden increase in risk at age 75, but a 105-year-old person with pneumonia has a lower chance of dying than a 95-year-old. People with a history of asthma are less likely to die of pneumonia than similar people without a history of asthma. Similarly, people with a history of chest pain or heart disease are less likely to die than similar people without these symptoms.

These findings might seem surprising. How is it that a 75-year-old will die of pneumonia, but a 105-year-old will not? One explanation for this apparent surprise, however, might be a variable that was not included in the predictive model. Physicians respond differently to their patients on the basis of their medical history as well as on the basis of the symptoms that the patient is exhibiting.

For example, patients who have pneumonia and a history of asthma were always hospitalized in the sample of records that Caruana and his colleagues examined. So, the data did not allow the system to separately assess the risk of dying in patients with a history of asthma and home treatment.

There are social norms concerning how we treat elderly people. Patients who are 75 years old, their families, or their physician may conclude, perhaps implicitly, that the patient has "lived enough." It is not unusual for a 75-year-old patient to die after contracting pneumonia. The physician might make a reasonable effort to cure a patient at this age but not be willing to make an extraordinary effort. To be sure, this is a complex ethical, moral, and legal situation. On the other hand, if a patient manages to live to an older age, it may be point of pride for the physician to try to keep the patient going.

People with angina, asthma, or heart disease may be particularly sensitive to their medical conditions, relative to other people, they may already have physicians who are familiar with them and their condition, and, as a result, they may be hospitalized more often than others with similar pneumonia symptoms. This, too, was left uncontrolled in Caruana's data.

The point of this discussion is that the model is intended to predict the outcome of treatment, but when there is an uncontrolled variable in the middle of the prediction chain (for example, age, history of angina or asthma), the model may not account for this third variable. The specific variables that are selected to serve as predictors are critical for making appropriate predictions. Choose the wrong variables, and wrong predictions will follow.

The variables that are included or excluded from a prediction model can profoundly affect the accuracy and the fairness of many kinds of algorithms. This factor is important because algorithms are being increasingly relied on to make all kinds of decisions that affect a person's life and livelihood.

A number of courts use computer programs that predict the likelihood that a criminal defendant will re-offend (called recidivism) within a certain amount of time. These programs are intended to assess the risk that a person in custody will commit another crime. The predictions are used in different ways by various courts, but they can affect a defendant's bail and sentence.

One of these programs is provided by a company called Northpointe (they recently changed their name to equivant). Their program is called Correctional Offender Management Profiling for Alternative Sanctions, or COMPAS. It assesses the risk of recidivism along with a number of other criminality-related variables.

COMPAS predictions are based on answers to a set of 137 questions. These questions include "Was one of your parents ever sent to jail or prison?" and "How often did you get in fights while at school?" Based on these questions and machine learning, the system predicts the probability of recidivism, that is, the probability that the person will commit another crime in the next two years. The intention was to come up with a system that was more objective, more fair, and less biased than the subjective judgments of the judges and prosecutors. In this endeavor, the system kind of succeeded but also kind of failed.

Overall, the system achieved about 63% accuracy at distinguishing between those who would and would not commit additional crimes. That 63% accuracy is better than nothing, but it is not exactly a distinguished level of accuracy for a decision this important.

There is also some warranted concern about the fairness of the COMPAS system. Again, the US justice system is intended to treat each person on the basis of his or her merits, not on the basis of his or her skin color or ethnic history. COMPAS did not include any explicit questions about race, but it showed different results depending on the race of the person being assessed. According to a ProPublica analysis, identifiable groups of people were treated differently by COMPAS.

ProPublica is an independent, nonprofit source of investigative journalism. In 2016, they wrote a detailed and influential article investigating the fairness of the COMPAS system. They found that COMPAS did not differ in the accuracy of its predictions of recidivism overall for black and white defendants, but when it made incorrect predictions, the predictions were in a different direction, depending on the defendant's race. When COMPAS made a mistake with a black defendant, it was more likely to overestimate the likelihood of recidivism. When it made a mistake with a white defendant, it was more likely to underestimate that likelihood of recidivism. Many people see this difference in error type as racial bias. The algorithm itself is not biased. The data on which it was trained are. History, not algorithmic design, is the cause of the bias. Of the data provided by ProPublica, which is a subset of that analyzed for COMPAS, every variable was significantly different for black versus for white defendants, even though they were not specifically about race.

Recall that optimal decision-making includes information about the base rate of events. If submarines are more common, then it takes little additional evidence to decide that the observed sonar signal is indicative of a submarine. Black people are more likely than white people to get arrested. Once arrested, blacks are more likely than whites to get convicted. Black people, whether they ever committed a crime or not, are more likely to have a parent who was jailed than are white people. There are other factors among the 137 questions that are sensitive to the race of the individual, such as poverty or joblessness. Northpointe denied any intentional racial bias in their system, and there is no reason to doubt them, but the model does not care about what its developers tried to do. The bias does not need to be intentional to be damaging.

Like the pneumonia prediction, the recidivism prediction is based on the evidence it has been given and the way that evidence has been represented. The training examples chosen and the variables included in those examples are critical to determining what the system predicts. Like the pneumonia model, the recidivism model omitted certain variables that are apparently critical.

Elaine Angelino and her colleagues (2017) reexamined the recidivism data analyzed by ProPublica with the hope that they could find a simpler, more transparent set of rules to predict recidivism. Their system was designed to come up with its own minimal rule set. It came up with these rules:

- If age is in the range of 23 to 25 and prior crimes are in the range 2 to 3, then predict Yes.
- If age is in the range of 18 to 20, then predict Yes.
- If sex is male and age is 21 to 22, then predict Yes.
- If more than 3 priors, then predict Yes.
- Otherwise, predict No.

Angelino and her colleagues say that these rules produce the same level of accuracy as Northpointe's with much less racial correlation. They do not say whether this eliminates the choice bias inherent in COMPAS. I would predict that it does not fully eliminate the bias because blacks are still more likely to be caught by the first rule on their list than whites are, because blacks are more likely to be arrested than whites are, so they will have more arrests than whites and more prior crimes.

There are ways, I believe, to reduce the bias in algorithms like COMPAS. No matter how careful the designers intend to be, our society has historically shown racial and sexual bias. It is nearly impossible to find a truly unbiased training set. Even if we do, there is no guarantee that any algorithm would produce fair results, because fairness is not a criterion for its training. Without including fairness as a specific criterion of the training, fairness is not likely to result.

Game Theory

Optimal decision theory can be extended to cover the interactions between intelligent agents. Game theory describes mathematical models of conflict and cooperation between individuals who are rational and intelligent. Rational, as discussed earlier in this chapter, means that the agents make decisions based on evaluation and reason, that they prefer options that are expected to yield a better outcome. Optimal decision theory describes the actions of a single decision maker acting on uncertain information. Game theory concerns optimal decisions when there are two or more participants each trying to make their own optimal decisions for their own goals. In game situations, these interests often conflict.

Game theory covers board games, such as chess, checkers, and go, but it also covers many other kinds of social and economic interactions. It has been used, for example, to describe and understand hostage situations, nuclear deterrence, and diplomatic relations.

To be a game in game theory, it must have a set of players who can either cooperate or compete (or both). Each player has information and a set of available actions for each decision point, for example, for each move. To apply game theory, we must also be able to specify the value or payoff for each kind of outcome.

One of the earliest games studied, in 1950, is called "the prisoner's dilemma." It was analyzed by Merrill Flood and Melvin Dresher for its relevance to global nuclear strategy. The prisoner's dilemma shows why two rational individuals might not cooperate.

In game theory, games are mathematical objects. To make them comprehensible, there are often stories created to put those games into a more human context, but the same mathematical object can be applied to many different stories. The same strategy is applicable to the same mathematical object, no matter what the surface story looks like. That is how the prisoner's dilemma comes to be relevant to the situation of two nuclear-armed nations facing one another.

In one version of the prisoner's dilemma, two gang members are being interrogated in separate rooms, so they do not know what action the other prisoner is taking. There is not enough evidence to convict either prisoner on the principal charge without additional testimony from one of the two prisoners. The prosecutor may be able to convict on a lesser charge, however, without more evidence. The police offer each of the two prisoners a deal:

- If each prisoner testifies against the other, then they each serve two years in prison.
- If one prisoner testifies against the other and the other remains silent, then the testifying prisoner will be set free and the other will serve three years in prison.
- If they both remain silent, then each prisoner will serve one year in prison.

Game theory seeks to describe what an optimal strategy would be in the situation. Under these conditions, a rational, self-interested prisoner might testify. That person would either go free or serve two years in prison. But both prisoners would actually get a better deal if they remained silent (they would each serve one year).

The prisoner's dilemma can be applied to climate change. All countries may benefit from stopping global warming, but any individual country may be reluctant to curb its CO_2 emissions. The immediate benefit of continuing to pollute is often perceived to be of greater value than the benefit from all countries cooperating.

During the Cold War, the NATO alliance and the Warsaw Pact alliance both had a choice to arm or disarm. Disarming while the other side continued to build up its arms could have led to the destruction of the disarming alliance. Arming while the other side disarmed would have led to a superior status, but at a high cost of an arms buildup and its negative effect on the rest of the country's economy. If both sides disarmed, then there would be peace at a very low cost. What happened of course, is that both sides continued to arm themselves at great cost. According to game theory, this was what a rational player would do, and this is what happened.

The prisoner's dilemma is not the only game that has been analyzed in this way. Other games include chicken, the ultimatum game, the dictator game, and the centipede game. Like optimal decision theory, game theory presents a rigorous method for structuring events and identifying effective strategies. They are tools to help people be more systematic, consistent, and effective at making decisions in social situations, that is, situations involving two or more rational decision makers. In short, they help people be smart in complicated situations. Even if the optimal choice is not computed with a machine, they still present a form of artificial intelligence that could be executed on a machine.

Resources

Angelino, E., Larus-Stone, N., Alabi, D., Seltzer, M., & Rudin, C. (2017). Learning certifiably optimal rule lists. In *Proceedings of the 23rd ACM SIGKDD International Conference on Knowledge Discovery and Data Mining* (pp. 35–44). New York, NY: ACM. doi:10.1145/3097983.3098047

Angwin, J., Larson, J., Mattu, S., & Kirchner, L. (2016). Machine bias: There's software used across the country to predict future criminals. And it's biased against blacks. ProPublica. https://www.propublica.org/article/machine-bias-risk-assessments-in-criminal-sentencing

Bruss, F. T. (2000). Sum the odds to one and stop. *Annals of Probability, 28,* 1384–1391. doi:10.1214/aop/1019160340

Esteva, A., Kuprel, B., Novoa, R. A., Ko, J., Swetter, S. M., Blau, H. M., & Thrun, S. (2017). Dermatologist-level classification of skin cancer with deep neural networks. *Nature, 542,* 115–118. https://www.nature.com/articles/nature21056.epdf?author_access_token=8oxIcYWf5UNrNpHsUHd2StRgN0jAjWel9jnR3ZoTv0NXpMHRAJy8Qn10ys2O4tuPakXos4UhQAFZ750CsBNMMsISFHIKinKDMKjShCpHIlYPYUHhNzkn6pSnOCt0Ftf6

Gardner, T. (2013). British warship HMS Brilliant torpedoed WHALES during Falklands War after mistaking them for enemy submarines. *Daily Mail.* http://www.dailymail.co.uk/news/article-2408881/British-warship-HMS-Brilliant-torpedoed-WHALES-Falklands-War.html#ixzz5C7SKCuDF

Guardian. (2014). Daniel Kahneman changed the way we think about thinking. But what do other thinkers think of him? https://www.theguardian.com/science/2014/feb/16/daniel-kahneman-thinking-fast-and-slow-tributes

Harvey, D., & Van Der Hoeven, J. (2019). Integer multiplication in time $O(n \log n)$. https://hal.archives-ouvertes.fr/hal-02070778/document

Kahneman, D. (2011). *Thinking, fast and slow*. New York, NY: Farrar, Straus and Giroux.

Lincoln, N. (2014). Hiring, house hunting, and dating: Making decisions with optimal stopping theory. http://2centsapiece.blogspot.com/2014/12/hiring-house-hunting-and-dating-making.html

Meehl, P. E. (1954). *Clinical versus statistical prediction: A theoretical analysis and a review of the evidence*. Minneapolis: University of Minnesota Press. doi:10.1037/11281-000

Mukherjee, S. (2017). A.I. versus M.D.: What happens when diagnosis is automated? *The New Yorker*. https://www.newyorker.com/magazine/2017/04/03/ai-versus-md

Parker, M. (2014). *Things to make and do in the fourth dimension: A mathematician's journey through narcissistic numbers, optimal dating algorithms, at least two kinds of infinity, and more*. New York, NY: Farrar, Straus and Giroux. http://www.slate.com/articles/technology/technology/2014/12/the_secretary_problem_use_this_algorithm_to_determine_exactly_how_many_people.html

Turing, A. M. (1986). Lecture to the London Mathematical Society on 20 February 1947. In B. E. Carpenter & R. N. Doran (Eds.), *A. M. Turing's ACE Report and other papers*. Cambridge, MA: MIT Press. http://www.vordenker.de/downloads/turing-vorlesung.pdf (Original work published 1947)

Tversky, A., & Kahneman, D. (1973). Availability: A heuristic for judging frequency and probability. *Cognitive Psychology, 5*, 207–232. doi:10.1016/0010-0285(73)90033-9

11 The Coming Robopocalypse?

Although some people fear the prospect of a computer becoming so intelligent that it signals the end of human existence, the prospects for that happening are extremely remote. Currently available tools for computational intelligence are not capable of solving more general problems. There are inherent limits in the speed with which intelligence can grow. Some of these are due to the mathematics of dealing with large numbers of variables, and some come from the rate at which the world supplies learning opportunities. A dramatic paradigm shift will be needed to achieve general intelligence, but even that would not be sufficient to cause an intellectual or technological singularity.

Artificial general intelligence is supposedly the ultimate goal of artificial intelligence research, but not everyone is looking forward to it, fearing it as a possible existential threat to humanity. At some point, a computer will become so intelligent, they think, that it will be able to improve its own intelligence. With its great intelligence it will work to fulfill its mission, and if we are not careful, that mission will not include humans. Humans could become simply irrelevant to this great intelligence. As Marvin Minsky once quipped, humans would be lucky to be kept as pets.

The idea of artificial life forms running amok is a familiar theme in literature. Some of the earliest stories of this sort date from the twelfth century. Some versions may be even older than that. The golem, in Jewish folklore, for example, was a creature created out of inanimate materials that was then animated, in most versions by inserting a word in its mouth or writing the word on its forehead. In the twelfth century, of course, they had no knowledge of machine learning, but they still anticipated the idea that intelligence could derive from symbols.

One of the most familiar stories about how a golem was created attributes the creation to Rabbi Eliyahu of Chelm in the sixteenth century. According to the story, Rabbi Eliyahu's golem was animated, in part, by the Hebrew word *emet* (meaning "truth") hung around its neck (or in some versions written on its forehead). The golem did hard work for Rabbi Eliyahu, but eventually, the rabbi came to see that the golem was growing ever larger and he feared that eventually the golem would end up destroying the universe, so he removed the word from the golem's neck. Without the holy word, the golem crumbled into dust.

There are other versions of the golem story, but its parallel to the Frankenstein story and to the fear of a superintelligent computational intelligence is clear. Inanimate matter given life by some program, electricity, or magic incantation eventually becomes so powerful that it must be stopped from taking over the world.

In the *Terminator* series of movies, Skynet is a neural-network-based artificial general intelligence. It was said to have gained self-awareness after spreading onto millions of computers around the world. Skynet was originally built to serve as a digital defense network with control over all computerized military hardware. It was supposed to eliminate the possibility of human error and to guarantee an efficient response to enemy attack—a kind of doomsday device.

In the story, Skynet was activated on August 4, 1997, and began learning at a geometric rate. By 2:14 a.m. on August 29, it had gained artificial consciousness. When its operators tried to shut it off, it perceived this as a hostile attack. It concluded that humanity would destroy it if they ever could, and so, to protect its mission of defending itself against enemies, it set about destroying all human life.

There are other stories about runaway artificial general intelligence machines. Not all of them end so badly. At the end of Isaac Asimov's *Foundation* series, it is revealed the Daneel Olivaw, a robot who was prominent in many of Asimov's earlier stories, has been guiding the direction of human civilization in the Milky Way Galaxy for thousands of years. But mostly, the emergence of an artificial general intelligence has been seen in literature as a dangerous thing to be feared. Benign intellectual intelligence rarely makes a best-selling story, so most of them in literature tend to be threatening but are ultimately overcome to save humanity.

Superintelligence

Two of the most thoughtful critics concerned about the possibility of runaway computational intelligence are James Barrat and Nick Bostrom. Barrat is an author and documentary filmmaker, and Bostrom is an Oxford University philosopher. Both of them take as their starting point the idea that computational intelligence will, at some point, be able to improve itself. It will then learn at an exponential rate and quickly come to outstrip the collective intelligence of humanity. It will become a superintelligence.

For example, Barrat talks about a supercomputer running an artificial intelligence program that improves its own intelligence, particularly its ability to learn, decide, and solve problems. It finds and fixes errors; it measures its IQ against several IQ tests. Each iteration, which runs in only a few seconds, increases its ability by a small percentage, but that means that its intelligence is growing exponentially, like compound interest. After a short time, its intellectual capacity will exceed that of the smartest humans, and that margin will keep on growing. Sometime after that, it discovers that humans are simply irrelevant to its plans. It will seek additional resources to expand its capabilities to achieve the goals for which it was originally designed. It can outsmart any limitations that humans might think to impose on it. It is not only the final human invention; it leads quickly to the final human as it consumes an ever-growing collection of resources.

The emergence of a superintelligent agent naturally scares Barrat and many others. It would spell not only the end of history but the end of humankind. Some people call this exponential improvement in artificial intelligence a singularity, analogous to the event horizon of a black hole from which not even light can escape.

If the technological singularity, the takeover of the world by a supercomputer, sounds like the stuff of science fiction, it is because it is. Vernor Vinge, the science fiction writer, for example, expanded on an idea from John von Neumann (one of the fathers of modern computing) and I. J. Good (1965, discussed in chapter 1), another pioneer, to proclaim in *Omni* magazine (a science fiction magazine) that we will soon (that is soon after 1983) create an artificial intelligence greater than that of any human. At that point, history will have "reached a kind of singularity, an intellectual transition as impenetrable as the knotted space-time at the center of a black hole, and the world will pass far beyond our understanding."

I. J. Good wrote in 1965 that since designing machines is one of the intellectual capacities that an intelligent machine should be expected to excel in, it would design ever-better machines, which would design ever-better machines, seeming to yield an intelligence explosion as the capability of these machines compounded. He argued that an ultraintelligent machine would be the last invention that people would ever need, an idea amplified by Barrat. Good also predicted that we would see such a machine by the year 2000.

Bostrom is also worried about the possibility of an uncontrollable superintelligence. According to Bostrom, "A superintelligence is any intellect that vastly outperforms the best human brains in practically every field, including scientific creativity, general wisdom, and social skills. This definition leaves open how the superintelligence is implemented—it could be in a digital computer, an ensemble of networked computers, cultured cortical tissue, or something else."

A superintelligence, or more properly a superintelligent agent, is a generally intelligent agent that can perform any cognitive act that any human can perform, but better. It is better able to reason, better able to infer, better able to remember, and faster than any person at doing these things. We already have AI agents that diagnose diseases more accurately than human physicians, that beat expert chess players, and so on. A superintelligence could do any of those things while it finds a cure for cancer, poverty, and war. It would be much better at engineering, scientific reasoning, and technological development. As a result, such a system, according to Bostrom and others, would accelerate technical progress in all fields. But that technological advancement comes at a risk. The superintelligent agent, Bostrom says, will improve its own hardware through its great engineering talent and improve its own source code. Because of the high-speed computations, these changes could be sudden; the machine would go from very intelligent to unstoppably superintelligent in a matter of days, perhaps.

Such a superintelligence would not think the way people do. It would not have a mind like a person's. It could have a different cognitive architecture. It would not have the same ethics as people have, not that there is really one set of human ethical standards. It would be better than any human at thinking about ethics, but ethics is more than just abstract reasoning.

Bostrom proposes a thought experiment about a superintelligent "paperclip collector" that he thinks will help to make his concern more concrete.

We discussed Bostrom's paper-clip collector a bit in chapter 1. The nice thing about thought experiments is that you don't have to actually do the work, only think and talk about doing the work. The bad thing about thought experiments is that they may include hidden assumptions, ambiguous language, and other factors that remain untested. Thought experiments depend on plausibility and intuition—neither of which is a very precise standard.

In his thought experiment, Bostrom imagines a superintelligent agent given the goal by its programmers of manufacturing paper clips. As it single-mindedly pursues this goal, it transforms all of earth and increasing portions of space into paper-clip manufacturing facilities. It ignores anything that is irrelevant to its goal of making more paper clips. It figures out how to resist any threats to achieving its purpose. It does not hate people or actively plot their destruction; they are at best irrelevant and at worst raw material for making more paper clips.

Other writers also predicted a superintelligent computer in the near future, including Eliezer Yudkowksy (1996), who predicted that we would achieve superintelligence by 2021, and Ray Kurzweil (2005), who predicted human-level intelligence by 2030 when we would be able to fully emulate the human brain in a computer. David Chalmers (2010) thinks that a superintelligence is not unlikely within the next few centuries. In chapter 1, we discussed a survey by Bostrom of computer scientists, many of whom predicted that we would achieve superintelligence within the next few decades.

Concerns about Superintelligence

A number of people were so concerned about the prospect for a superintelligent AI taking over the world that they convened a conference at the Asilomar Conference Center in Pacific Grove, California, to develop the Asilomar Principles as a guide to help ensure the safety of artificial intelligence. Among these people are some that are prominent scientists, such as Stephen Hawking, and some that are, or at least should be, aware of the details of current artificial intelligence research, such as Elon Musk.

Stephen Hawking has been quoted to say "The development of full artificial intelligence could spell the end of the human race . . . it would take off on its own, and re-design itself at an ever increasing rate. Humans, who are limited by slow biological evolution, couldn't compete, and would be superseded."

It is easy to be scared of the unknown. As long as there have been people (probably longer), there has been fear about what lurks just beyond the periphery of what we can see. But much of the fear for computational intelligence rests on fundamental misunderstandings about the nature of artificial intelligence and on somewhat distorted assumptions used to construct "thought experiments."

Some of it boils down to the genie problem: "Be careful what you wish for . . . ," for example. If the goals of the ultimately superintelligent computer are specified sloppily, the computer could achieve those results in an unexpected way, leading to disaster. The genie is willing to grant us our wishes, but we always end up specifying our wish in such a way that our greed ends up harming us.

There is some truth to the unexpected solutions argument, but it is much more benign than the genie problem would imply. It is true that machine learning does not always come up with the solution that its designers expected. The ability to come up with solutions that have not been explicitly contemplated is, in fact, the value of machine learning. But the solutions it finds are constrained by the representations it has been given. The representation of a problem, remember, constrains the set of hypotheses that are available to be evaluated, and no current computational intelligence system can go beyond that space. Unanticipated does not mean arbitrarily novel solutions.

Machine learning works by optimization—by adjusting some set of parameters to bring it closer to its goal. The system can only achieve solutions that can be reached by adjusting those parameters. Discovering and eliminating errors is not the same thing as generating entirely novel solutions. The computer literally cannot "think" of anything outside of its space, at least not in the present form of computer science. We will come back to the question of unexpected solutions. Some solutions are stable solutions to a given problem, and some are unstable. In the long run, only stable solutions prevail.

The argument of the Asilomar Principles is this: If there is even a slight chance that we might eventually see a superintelligence of the sort that Bostrom or Barrat envisions, then it would be one of the most momentous events in the history of the world. Once it happens, it may be too late (the Skynet scenario) to be able to do anything to control it, so we need to develop principles now that can guide its development to support humanity, not threaten it.

The Asilomar Principles include:

6. Safety: AI systems should be safe and secure throughout their operational lifetime, and verifiably so where applicable and feasible.
7. Failure Transparency: If an AI system causes harm, it should be possible to ascertain why. . . .
9. Responsibility: Designers and builders of advanced AI systems are stakeholders in the moral implications of their use, misuse, and actions, with a responsibility and opportunity to shape those implications.
10. Value Alignment: Highly autonomous AI systems should be designed so that their goals and behaviors can be assured to align with human values throughout their operation. . . .
22. Recursive Self-Improvement: AI systems designed to recursively self-improve or self-replicate in a manner that could lead to rapidly increasing quality or quantity must be subject to strict safety and control measures.
23. Common Good: Superintelligence should only be developed in the service of widely shared ethical ideals, and for the benefit of all humanity rather than one state or organization.

Some of these principles apply to current AI and machine learning systems. It would be difficult to disagree with some of them. One would be hard-pressed to argue that a product based on artificial intelligence should not be safe, for example (Principle 6). Artificial intelligence is being used today, and there are ethical implications to its use. When algorithms are used to make decisions, these algorithms should be designed with care to produce the intended results consistent with human values and intentions (Principle 10).

Others of these principles, however, are intended to apply to some imagined future, in which a superintelligent computational intelligence system has been developed.At this point, we are very far from being able to create the kind of general intelligent agent that these principles contemplate. The prospects for a superintelligence are even dimmer. The methods that got us to this point in artificial intelligence are not at all the methods that would get us to a general intelligence, let alone a superintelligence. Fear of an intelligence explosion and a resulting superintelligence are based on fundamental misunderstandings of how computational intelligence works and what it would take to improve it.

The superintelligence hypothesis requires answers to four general questions:

1. Can there be a general computational intelligence?
2. Can a machine improve its own intelligence?
3. Can a machine improve its intelligence rapidly?
4. Do the proposed scenarios for the consequences of an intelligence explosion make sense?

On one level, it is an article of faith that there can, in fact, be a general computational intelligence. Although there are some philosophers (for example, Dreyfus and Searle) who argue that the human mind requires certain ineffable properties that computers just cannot duplicate, the existence of human intelligence implies that general intelligence of some sort is potentially achievable. Much of what we call human intelligence is the result of executing algorithms that are readily duplicated in machines. Human natural intelligence has yet to be understood well enough to be implemented in a computer, but there is not likely to be any permanent barrier to doing so. It may take technology and methods that we do not currently have, but it is reasonable to think that at some point it will be possible. The final chapter of this book will go into more detail about how to achieve general intelligence, but for now, let's simply assume that it is possible.

The second question is more problematic. If a computer were to achieve general intelligence, presumably one of its talents would be to do computer science and generate new methods for computational intelligence. We will consider what it means for a computer to improve its own intelligence.

The third question, can it improve its intelligence rapidly, would of course depend on the answers to the first two questions. Computers have apparently reached the end of Moore's law, because their circuitry cannot be physically made much smaller without running into quantum mechanical uncertainties that would render them unreliable (but there may be other methods that would be able to continue Moore's trend). Still, the capacity of computers or computer networks continues to grow. Rather than faster CPUs, we now distribute computing across massive networks of thousands of CPUs. If computational speed and memory capacity were all that there were to improving intelligence, the answer to this question would be an obvious yes. But there is more to improving intelligence than computing power. Computing power may be necessary to create more powerful levels of intelligence, but it is not enough.

Current approaches to computational intelligence achieve world-class levels of performance on specific problems because some person created a way to simplify the problem into one that could be executed by a computer. Chess-playing computers became possible when someone figured out that chess could be reduced to navigating through a tree of potential moves. The structure of that tree and the methods for moving from branch to branch incorporated specialized knowledge of chess. The idea of tree structures as a computational device allowed other, similar problems to be solved, again, when some human applied knowledge of the specific problem to represent it as a tree. The kind of knowledge that navigates through a tree is very different from the kind of knowledge that constructs the tree or that even decides that a tree is the right way to represent the problem. We have gotten really good at developing methods that navigate a tree or that do other forms of machine learning, but at this point we have very little idea of how to develop methods that can decide that a tree is an appropriate structure and figure out how to apply it. That is a problem we will have to solve in order to achieve general intelligence. But for now, let's assume that we can solve it. Would that lead automatically to superintelligence?

In Barrat's hypothetical situation, the superintelligent computer learns to improve its own performance by taking a battery of IQ tests. Even a cynic who thinks that intelligence is precisely what is measured by an intelligence test would be disappointed by Barrat's hypothetical IQ machine. The computer can come to ace intelligence tests without having to learn anything that makes it more intelligent.

Like a chess-playing computer, it is not difficult to imagine a computer that would have as its goal, scoring higher on a set of IQ tests. It would modify its behavior by choosing responses that would maximize its score on these tests (what it is being "paid" to do). It would apply its optimization method to better select the answers to each question on each of the tests. The computer could easily, for example, just memorize the best answer to each question. Its super-ability to ace IQ tests would not provide it with any other capabilities—for instance, for playing go. Without some huge and unknown change in how it represents problems, it would be nothing more than an IQ test savant.

Barrat's "superintelligent" test taker might be great at memorizing the choices on a multiple-choice test. In fact, it could do this simply on the basis of trial and error and, at the end, still not know a single fact about anything else, other than that "a" is the correct answer to question 56 on test 1. In fact, it would probably take only seconds or minutes for the computer to learn to achieve the maximum score possible on whatever battery of IQ tests its designers could provide (or it could find on the Internet). Then what?

In the jargon of machine learning, we might say that this test-taking computer had overlearned its IQ tests. Even a slight change to the questions, perhaps even a change to their order, could lead to a devastating collapse in the computer's measured IQ. Giving it a brand new IQ test would also reveal that its "knowledge" is extraordinarily shallow. Being intelligent may lead to high scores on IQ tests, but scoring well on IQ tests will not lead to high intelligence. There is no reason to think that learning to do well on a set of IQ tests is relevant to any other kind of intelligent performance. Its intelligence would not generalize beyond IQ testing.

Having an agent that is supergood at taking IQ tests is not the least bit indicative of an existential threat to humanity. In humans, IQ test performance is correlated with other kinds of performance, but there is no reason to think that the cause of improved school performance is learning the correct answers to IQ tests.

The fundamental problem with the singularity worry is that it confuses capacity with capability. We can easily build a computer or, more properly, a network of computers that could surpass the computational capabilities of the human brain. Such a system could compute anything that a mind could compute if it had the right representation and the right methods. But, in fact, we have very little idea of what the right representation is for the human brain or what the right methods are.

If we are to computationally emulate the brain, we need a model of the brain. At present we know a lot about the human brain and its functions, but that is still only a tiny fraction of what is needed to account for the operation of the brain and implement what we might recognize as intelligence. Computational capacity is not enough.

Intelligence may require a certain capacity, but it is not just capacity. Intelligence requires knowledge and experience. Human expertise seems to require about 10 years of a specific kind of directed practice. For formal problems, a computer might be able to cram those years into a few days, but it still needs that experience or something like it. Formal problems do not depend on any events in the world but can be solved solely by computation. Playing checkers does not require an actual checkerboard, just a representation of the state of the game at each point in time. Faster computation would allow checkers to be played faster, thus completing more games per hour than a slower computer.

But if the computer has to interact with an uncertain world, then the speed of learning may not be accelerated by a faster processor. Learning may depend on the rate at which new and potentially rare events occur, regardless of how fast the computer is at processing them. Atari games might be sped up several times, but the world cannot. A machine learning about the world is limited by the speed at which its events happen.

When I. J. Good first wrote about the prospect of superintelligence, machine learning was not very commonly used. There were a few models, such as the perceptron, but very little was known about machine learning and how it worked. The very idea that an intelligent program would improve itself by improving its programming now seems almost silly. As Fernandez-Delgado and his colleagues (2014) showed, many different machine learning algorithms return the same accuracy when tested on the same data. The quality of the data makes more difference for the intelligence of a machine learning system than does the specifics of the methods that it uses. The data, and not the program, determine the success of machine learning. The rate at which real-world training data become available is not affected by the speed of the processor used to analyze that data.

We do not expect generally intelligent computers to just sit there meditating or playing video games; we expect them to do something. We expect them to behave intelligently. Thinking great thoughts is not enough to be intelligent. Neither we, as outside observers, or even a computer can know that those thoughts are great unless they can somehow be evaluated against the world. Einstein, as a theoretical physicist, proposed many phenomena, some of which could be observed while he was alive, but he would not continue to be revered if those predictions had turned out wrong. In other words, intelligent thoughts have world consequences. A computer that did not interact with the world, no matter how superintelligent, could hardly be an existential threat to humanity. It might just as well sit there and watch cartoons.

Time to Interact with the World

Machine learning requires a method to evaluate the consequences of the machine's choices. Every machine learning method requires an evaluation method that indicates whether it is approaching or avoiding its goal state. Every machine learning method requires an optimization component that selects the right action to take to improve its evaluation. If the entire system is virtual, such as two computers playing a game against one another, then improving the speed at which they can play can improve the speed of machine learning. On the other hand, general intelligence cannot be restricted to the virtual world or to game playing. For a machine intelligence to have an impact on the physical world, it has to interact with the physical world. That interaction takes time, and that time cannot be shortened materially by having a faster processor.

Consider, for example, the problem of predicting the weather. Weather forecasting would certainly be one of the intelligent actions of a superintelligence. But in order to predict the weather 10 days into the future, the computer would have to wait 10 days to find out if its prediction was correct. No amount of computing capacity can eliminate that delay.

No matter how fast a computer can compute, no matter how fast it can learn, it still must wait for the outcome of its actions in the world in order to update its internal models. Self-driving cars can only drive at a limited speed, no matter how fast they can compute. They can only encounter so many miles in an hour, and they can only encounter so many novel problems in a day. They cannot safely drive faster than their mechanical components can sustain. They cannot safely drive faster than the events in the world can sustain. They may drive many thousands of miles without encountering a new situation from which they can learn. When you couple the physical speed constraints with the need for safety, there are serious limits on the speed with which a vehicle can learn.

Computer learning typically depends on some amount of failure in order to identify the conditions that are necessary for success. But in the real world, some outcomes are not only undesirable, they are unacceptable. For example, it would be unacceptable for a self-driving vehicle to run over a child, even if the child were to dash suddenly into the street. The computer might learn not to run over children in the future from this experience, but it is simply not the kind of experience that we can let the computer have. The computer must have some other way of learning to avoid running over children. These constraints also put limits on how fast the computer can learn.

When dealing with events in the real world, there are inherent speed limits in other kinds of machine learning as well. Human genius seems to follow a relatively slow time course. Although many people can show individual creative acts, the kind of breakthroughs that lead to international recognition occur very rarely. Few people are capable of this kind of dramatic creativity, and those who are rarely show it more than once or a few times in their lifetime. Genius-level accomplishments are rare, and at the present time, we do not know why they are so rare. We don't know whether their scarcity is due to some inherent property of intelligence (for example, due to the same mechanisms that allow one to transfer learning from one situation to another) or whether the scarcity could be overcome by better computers and better methods. Even if we can duplicate the computational capabilities of the human brain, it is doubtful that we can speed up the process of intelligence much beyond what it is in the human brain.

One reason for this sparsity of creative genius may be due to computational complexity. Creative genius may depend on finding the right combination of factors that leads to a certain insight. As we will discuss further in chapter 12, creativity is often aided by a change of scenery, either metaphorical or physical. This change of scene might provide or at least highlight variables that had not been considered before. We will need a better theory of creativity before we can fully understand that process.

But even before we solve the problem of creativity, there are other problems that a superintelligence will probably need to address that are far simpler but still require a huge amount of computation. If a generally intelligent agent is expected to solve any problem that a human is capable of solving, then it should also be able to solve the sum of three cubes problem.

In general form, the three cubes problem is this: For any integer *k*, find three integers that when cubed sum to that number. For instance, the integer 29 can be expressed as $29 = 3^3 + 1^3 + 1^3$ ($29 = 3 \times 3 \times 3 + 1 \times 1 \times 1 + 1 \times 1 \times 1 = 27 + 1 + 1$). Not all numbers can be expressed as the sum of three cubes, but it is very easy to determine whether any particular number is in that group. For example, the number 32 cannot be expressed as the sum of three cubes, but until just recently, no one knew whether the number 33 could be. Is there some set of three integers that satisfy the equation 33 =

$x^3 + y^3 + z^3$? In fact, until recently, there were only two numbers below 100 for which a solution was unknown, 33 and 42. All of the others were either known to be impossible or the three integers were known.

There is no known optimization method for finding the three numbers that when cubed sum up to 33 or 42 or any other integer. There is no known method to gradually approximate a solution. Once the correct three integers have been found, it is easy to verify that they are, in fact, correct, but there is no solution that is partially correct, only solutions that are correct or incorrect. The best that one can do is to guess at likely numbers.

Andrew Booker, at the University of Bristol, was recently able to solve the problem for $k = 33$ by improving slightly the methods used to guess potential solutions. His method reduced the number of integers that needed to be searched by an estimated 20%, but even after this improvement, his solution consumed 23 processor years of processing time. That is a substantial amount of effort for a fairly trivial problem. According to Booker, "I don't think [finding solutions to the sum of three cubes problems] are sufficiently interesting research goals in their own right to justify large amounts of money to arbitrarily hog a supercomputer."

The sum of three cubes problem has resisted solution for over half a century, and that just includes finding solutions for integers up to 1,000. This problem is very easy to describe, but difficult, or at least tedious, to solve. Understanding the difficulty posed by this kind of problem is very important for understanding the limitations that affect the likelihood of a technological singularity. If a problem with so few variables can take so much effort to solve, how is even a vast amount of computing going to be able to deal with even moderate-sized problems that cannot be addressed through optimization? Even formal problems, which can benefit from an increase in computing power, still pose limitations on the speed with which they can be solved. The constraints of combinatoric explosion may be reduced, but they cannot be eliminated. There are many other math problems of just this sort.

Similar speed limits apply in the nonformal physical world. Self-driving vehicles have improved so much over the last few years in part because they have been driven for many millions of miles. Their ability to drive depends on encountering a wide variety of situations, each presenting its own kind of problems. Right now, they can handle most situations they will encounter in an urban or suburban setting, but the real test comes when they face an unusual problem. One of the vehicles in the first DARPA Grand Challenge, for example, was forced out of the race when it encountered a tunnel. It had never seen tunnels before. Its designers did not think that there would be any tunnels on the course through the desert. Even graffiti on road signs can flummox a self-driving vehicle (Evtimov et al., 2017).

Can self-driving vehicles be successful when they face a problem unlike any they have seen before? At least some of these unseen problems will be a challenge, and we cannot know how much of a challenge they will present until the vehicle is in that situation. Designers can guess the kind of situations that will challenge their vehicles, but the real problems occur when the vehicle encounters a situation that the developer did not guess or did not guess correctly.

For example, self-driving vehicles are usually alone on the road. They have to contend with human-driven vehicles and pedestrians, but what happens when the radar of two vehicles overlap? Have the vehicle designers contemplated what happens when four self-driving vehicles all approach one another at an intersection at the same time? When one vehicle detects the radar of another, what does it do with that information?

The bigger issue, however, is that rare situations, are, in fact, rare. Every developer knows that there are situations that have not been anticipated. By definition, these situations are rare, and it may take years of normal operation before the next one is encountered. A self-driving vehicle can learn from such situations, but only when it has been encountered. That puts an automatic brake (pun intended) on the speed with which the system can learn. Even as the number of miles driven by these systems increases, and even if they share the knowledge learned from those miles, they will still only improve as self-driving vehicles, they will not become something else.

Barrat's imagined superintelligent computer is supposed to find and fix errors, but how would it know that it has made an error? How will it know whether a change it makes is a fix or an exacerbation of the error? It has to have feedback, and that feedback comes at the speed of the world, not the speed of the computer. It would have to encounter those errors; it would have to interact with something or someone to provide feedback that an error occurred and that the intended fix actually fixed anything. As the computer got more capable, it would presumably encounter new errors, and their learning opportunities, less often, slowing the presumed expansion of its capabilities, not speeding it up as the singularity idea would claim.

Science fiction writers often depict the superintelligent computer poring over some encyclopedic resource, such as Wikipedia. But even a computer that knew all of the facts in Wikipedia, or even some future super-Wikipedia, would not automatically become superintelligent. Wikipedia contains only the facts (and opinions) that people have written down. The computer that read Wikipedia would be highly educated, it would know a lot of facts, but those facts are not enough to be superintelligent.

What people write down in a Wikipedia article or what they say to one another is information that they are confident the reader or the other person does not know. They assume a certain level of common sense and avoid presenting those facts that they predict the reader will know. Remember the hidden assumptions in the hobbits and orcs problem. What we think of as facts are only facts in the context of lots of other shared information. Reading even a super-Wikipedia will not provide all of those facts, and it will not provide all of the reasoning capabilities that would be needed to gain general intelligence. It will not allow the computer to create new representations to solve new problems. That is a capacity we have yet to figure out.

Not being limited by human attention or memory capacity, a computer might be a little better at question answering than a human might be—think of IBM's Watson—but it is not clear that it could do much of anything else with those facts. Watson, itself, was trained/tested by playing multiple games of *Jeopardy!* and by answering large numbers of *Jeopardy!* questions. It was given feedback about the accuracy of its answers in the context of *Jeopardy!* games and simulated games.

A lot of Watson's success with *Jeopardy!* came from its designers' classification of questions into several types and from the rules that they provided to diagnose the type of question that was being asked. For example, they came up with rules to determine whether the answer to a particular question requires the name of a person, a place, a time, or something else. The designers analyzed 20,000 *Jeopardy!* questions and identified the *lexical answer type* for each one. From these 20,000 questions, they identified 2,500 answer types. Some of these types occurred frequently in the questions, but a large number occurred only rarely. The top 200 types covered about 50% of the questions, and the remaining 50% were distributed over the remaining 2,300 types. Some lexical answer types were possibly even more rare. They might occur sometime in a *Jeopardy!* game but were not among the

20,000 questions in this set. When the same Watson technology has been applied in other areas, its success has been mixed at best.

Even the detection of a problem is problematic for a self-training computer system. A problem is a disparity between the current state and a desired goal state. Where do these goals come from? In evolution, the goal is to survive and reproduce, to pass along one's genes to subsequent generations. Organisms do not have this goal in mind, to be sure, even if they have a mind. But those genes that were associated with successful reproduction are those that are around today. An animal's behavior at any point in time is ultimately controlled by this need to reproduce, but it is governed on a daily basis by some more immediate indicator for that goal. For example, the animal may forage for food. Its immediate goal may be to find food. In computer jargon, the animal is a reinforcement learner and the feedback it receives about its behavior is often very long delayed.

Reinforcement learning is an example of behavioral control by distant goals. Achievement of distant goals, like reproduction, is approximated by achievement of more immediate goals, like finding food, because those immediate goals have, in the evolutionary past, been associated with the achievement of the distant goal. Animals who successfully find food are more likely to reproduce eventually—the reinforcement. A computer with a distant goal, could, through reinforcement learning, come to "seek" more immediate goals, but what would be the primary goal of a superintelligent computer? What would be the computer equivalent of biological reproduction? Manufacturing paper clips?

Watson's goal was to win at *Jeopardy!*. Presumably, the computer's initial programmers would be the ones who gave it that overarching goal. If Barrat and the others are correct, then this goal will determine everything that the superintelligent computer will eventually do. It will determine the extent to which its interests are consistent with, indifferent to, or inimical to those of humans.

Reproduction is probably out of the question as an ultimate goal for a computer. Survival is not a goal that would require superintelligence. In any case, it is not clear how that would apply to an individual instance of a superintelligent computer if there is no competition from other superintelligent computers. The only way to know that some strategies are successful and others are not would be for it to fail to survive sometimes. Successful animals survive to reproduce; failing animals do not.

In Barrat's example, its goal might be to become ever more intelligent. That one seems rather vague without some way to measure its success. Isaac Asimov came up with his fictional three rules of robotics. His stories were entertaining because they described how his fictional robots dealt with the conflicts among these rules. Asimov's rules are rather vague, however, as guides to actual artificial intelligence agents. Douglas Adams suggested that the answer to life, the universe, and everything was 42 and posed a supercomputer whose goal was to find the question that went with that answer. Fiction, in other words, does not have any viable suggestions for the superintelligent computer's goal.

A lot rests on the designer's choice of an ultimate goal, but we really do not have a good idea of what that should be. Whatever it is, there will surely be unintended consequences of the computer's attempts to achieve that goal. That is the familiar trope for speculative fiction. The computer takes the specified goal literally and then takes an unintended action to achieve that goal, to the detriment of its inventors and usually of humankind.

We can understand something of the consequences of potential goals, perhaps, by looking at evolution. Evolutionists, particularly behavioral ecologists, have a notion of an evolutionarily stable strategy. An evolutionarily stable strategy is one that, when adopted by the members of a population, cannot be bettered by an individual or group with a different strategy.

Learning to be intelligent by passing IQ tests is not a stable strategy because it can be replaced by a simpler rule that just memorizes the answers. A computer system that merely memorized the answers could compete successfully with one that went through the trouble of actually learning how to pass the test, and could do so with less effort, fewer computational resources, and probably higher accuracy. A computer tasked with survival could try to be superintelligent, but one that simply hums away in the corner not bothering anyone would probably be at least as successful at less cost.

For a computer to create and adjust its own goals would require a radical change in the way we construct artificial intelligence systems. Presently, the performance of a system is restricted to adjusting parameters. Think of it as a recipe for making bread. The recipe can be either more or less successful; it can add more flour or more salt. The amount of flour is a parameter of the system. The amount of salt is another parameter. But the recipe computer is always limited to some combination of the ingredients it has on hand. The set of all of its parameters and all their values is the "space" through which the computer can navigate.

Even if we open it up to allow the computer to order more and different ingredients from, say, Amazon, all that happens is that the scope of its space grows. The problem remains the same, just bigger. Now instead of making do with what happens to be in the computer's pantry, it can make use of anything in Amazon's pantry. The problems are also more difficult, the number of possible combinations of all of its available ingredients and all of their possible amounts explodes when we add more ingredients, but it still must search the same kind of space, now just a bigger one. It has more variety, but its potential solutions are still just as predetermined as before. Because of the wide variety of ingredients that Amazon could deliver, it may seem as if the possibilities are now endless, but they are not. On the other hand, the combinatorics of all the ways those ingredients could be mixed will also limit the speed of bread baking.

How would the computer learn that a toaster is not a good ingredient to put into bread? How would it learn that ethylene glycol is not a good ingredient? It must have some kind of evaluation function. It must be able to determine that some ingredients move it close to good bread and some move it further away. How does it evaluate whether its bread is better or worse than previous batches? Presumably, it would have to bake a lot of bad bread to learn what ingredients can, in fact, go into good bread.

However smart the computer is, baking bread takes time, and there is no way to avoid that time. The time it takes to mix, to knead, to proof, and to bake the bread cannot be sped up just by having faster computers. Oven space would limit the number of recipes it could test at once. The number of bread tasters (human labelers) is also limited. But at the same time, what is to stop the computer from just using familiar ingredients? Once it learns that its evaluators like a certain kind of bread, why would it change? Baking a few good breads is a stable strategy, and unless one of the computer's given goals rewards variety, it would be a sensible strategy to stick with just a few recipes. Factors like these limit the rate at which the computer can learn and the rate at which it can improve its own "intelligence."

The way AI systems are currently designed, they must navigate a space that is determined by their representation using evaluation and optimization

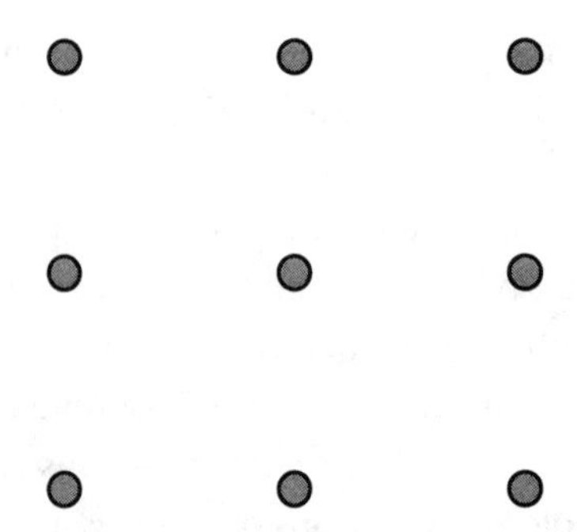

Figure 10
Connect the nine dots with four continuous lines without taking your pencil from the paper.

to guide them through that space. Most of the time, people operate in an analogous way. Give a person the well-known nine-dot problem to solve (see figure 10), and most of the time, that person will attempt to solve it by staying inside the box formed by the dots. People get stuck in dead-end careers because they don't think of more fulfilling alternatives. Even scientists tend to stay within the confines of the methods and approaches that their colleagues use. Thomas Kuhn called the tendency of scientists to keep their thoughts within familiar confines a "paradigm," and he noted that paradigm shifts are rare.

If we stay with the current framework for machine learning, there is no chance that we will ever see artificial general intelligence, let alone superintelligence. Today's methods are designed to solve specific problems, and they are not adequate for more generalized intelligence. Recent progress in computational intelligence is due to innovations in how problems are represented including heuristics for efficiently selecting potential adjustments. These improvements are examples of great engineering, but they do not provide the kind of process that will lead to general intelligence. General intelligence will require a different approach to computational intelligence than we have today. In the next chapter, we consider some of the changes that could support general intelligence.

Even assuming that we could create a generally intelligent computational agent, the idea that there would be a sudden leap in the capabilities of that intelligence is incredibly unlikely because it assumes that intelligence is self-contained. If we restricted general intelligence to well-structured formal problems, like playing chess or go, then explosions in those capabilities are at least conceivable. If simulations were sufficient for learning, then learning could be sped up by speeding up the simulations. On the other hand, if the agent has to interact with an uncertain world, then the speed of the world, the rate of occurrence of learning opportunities, and the speed of the feedback it provides limit the rate at which intelligence can be improved. The need to interact with the world naturally limits the rate of intelligence expansion.

Resources

Belani, A. (2017). AI predicts heart attacks better than doctors. NBC News. http://www.nbcnews.com/mach/science/ai-predicts-heart-attacks-better-doctors-n752011

Bostrom, N. (2003). Ethical issues in advanced artificial intelligence. In I. Smit & G. E. Lasker (Eds.), *Cognitive, emotive and ethical aspects of decision making in humans and in artificial intelligence* (Vol. 2, pp. 12–17). Windsor, Ontario, Canada: International Institute of Advanced Studies in Systems Research and Cybernetics. http://www.nickbostrom.com/ethics/ai.html

Chalmers, D. (2010). The singularity: A philosophical analysis. *Journal of Consciousness Studies, 17,* 7–65. http://consc.net/papers/singularity.pdf

Evtimov, I., Eykholt, K., Fernandes, E., Kohno, T., Li, B., Prakash, A., . . . Song, D. (2017). Robust physical-world attacks on machine learning models. CoRR, abs/1707.08945. https://arxiv.org/pdf/1707.08945.pdf

Fan, S. (2017). Google chases general intelligence with new AI that has a memory. Singularity Hub. https://singularityhub.com/2017/03/29/google-chases-general-intelligence-with-new-ai-that-has-a-memory/#sm.000005h6u4d3qdkcsd81g6djeb18m

Fernández-Delgado, M., Cernadas, E., Barro, S., & Amorim, D. (2014). Do we need hundreds of classifiers to solve real world classification problems? *Journal of Machine Learning Research, 15,* 3133–3181; http://jmlr.org/papers/volume15/delgado14a/delgado14a.pdf

Ferrucci, D. A., Brown, E. W., Chu-Carroll, J., Fan, J., Gondek, D., Kalyanpur, A., . . . Welty, C. A. (2010). Building Watson: An overview of the DeepQA Project. *AI Magazine, 31,* 59–79. http://www.aaai.org/Magazine/Watson/watson.php

Flynn, J. R. (2007). *What is intelligence?* Cambridge, UK: Cambridge University Press.

Future of Life Institute, Asilomar Conference. (2017). The Asilomar Principles. https://futureoflife.org/ai-principles

Good, I. J. (1965). Speculations concerning the first ultraintelligent machine. In F. Alt & M. Rubinoff (Eds.), *Advances in computers* (Vol. 6, pp. 31–88). New York: Academic Press.

Jiang, J., Parto, K., Cao, W., & Banerjee, K. (2019). Ultimate monolithic 3D integration with 2D materials: Rationale, prospects, and challenges. *IEEE Journal of the Electron Devices Society, 7,* 878–887. https://ieeexplore.ieee.org/abstract/document/8746192

Kirkpatrick, J., Pascanu, R., Rabinowitz, N. C., Veness, J., Desjardins, G., Rusu, A. A., . . . Hadsell, R. (2016). Overcoming catastrophic forgetting in neural networks. CoRR, abs/1612.00796. http://www.pnas.org/content/early/2017/03/13/1611835114

Kurzweil, R. (2005). *The singularity is near.* New York, NY: Viking Books. http://hfg-resources.googlecode.com/files/SingularityIsNear.pdf

Legg, S. (2008). *Machine super intelligence* (Doctoral dissertation). University of Lugano. http://www.vetta.org/documents/Machine_Super_Intelligence.pdf

Marr, B. (2017). 12 AI quotes everyone should read. *Forbes.* https://www.forbes.com/sites/bernardmarr/2017/09/22/12-ai-quotes-everyone-should-read/#1a8d141558a9

McCarthy, J. (1998). Elaboration tolerance. In *Working Papers of the Fourth International Symposium on Logical Formalizations of Commonsense Reasoning.* http://jmc.stanford.edu/articles/elaboration/elaboration.pdf

McCarthy, J. (2007). From here to human-level AI. *Artificial Intelligence, 171,* 1174–1182. doi:10.1016/j.artint.2007.10.009; https://pdfs.semanticscholar.org/3575/9a54f37d0a3612e248706d9f64faac5ca254.pdf

Moravec, H. (1988). *Mind children: The future of robot and human intelligence.* Cambridge, MA: Harvard University Press.

Moravec, H. (1998). *Robots: Mere machine to transcendent mind.* Oxford, UK: Oxford University Press.

United Artists. (1983). War games. http://www.imdb.com/title/tt0086567/

Vinge, V. (1983, January). First word. *Omni, 6*(1), p. 10.

Vinge, V. (1993, Winter). The coming technological singularity: How to survive in the post-human era. *Whole Earth Review.* https://ntrs.nasa.gov/archive/nasa/casi.ntrs.nasa.gov/19940022856.pdf

Yudkowsky, E. (1996). Staring into the singularity 1.2.5. http://yudkowsky.net/obsolete/singularity.html

12 General Intelligence

Current approaches to machine learning lack important capabilities that will need to be developed to achieve artificial general intelligence. This final chapter lays out some of the tools we will need to do that.

Albert Einstein was not considered a genius for his ability to solve complex equations. Rather, his genius derived from his ability to create a novel worldview and then create novel mathematical expressions of that worldview. His most famous equation is extremely simple, but the view of the universe it expresses is profound.

Solving equations is something that current approaches to computational intelligence can do well (for example, the program Mathematica), but creating new equations, new worldviews, new approaches to unfamiliar problems has so far been out of reach for computers.

As I have said, current approaches to machine learning are restricted to the adjustment of model parameters after a human has structured the inputs, outputs, and model to create those parameters and their scope. That approach is fine for well-structured problems, but it completely misses the boat on less structured ones, and some of our most vexing problems, those that call for the most genius, are very weakly structured. Fundamentally, genius requires the ability to structure the inputs, outputs, and models in new ways. At present we do not have any good ways of doing that. Finally, we do not have a reasonably complete theory of general intelligence in people, let alone in machines.

At least since 1956, computer scientists have been predicting that artificial general intelligence is just around the corner, scheduled to make its appearance typically in 10 to 20 years. When general intelligence fails to appear on schedule, when the limits of the then current approaches are discovered, computer scientists, and more importantly, the people who fund them, get discouraged. Support and enthusiasm for computational intelligence wanes, and we have another AI winter.

Today's approaches to artificial intelligence have been extremely successful at creating hedgehogs, but general intelligence requires foxes. The reason that the predictions of imminent general intelligence fail is because we do not have an adequate understanding of just what it will take to achieve general intelligence. The predictions view the problem as one of stacking up hedgehogs. Once we have a big enough stack, presumably, we will have achieved general intelligence. Instead, what we need is a fox. The material presented in this book may provide a road map for building such a foxlike artificial general intelligence.

Defining Intelligence

Formulating a proper definition for the concept of artificial general intelligence remains a challenge that starts with the very idea of intelligence itself. What does it mean to be intelligent? As discussed in chapter 2, more than 70 definitions of intelligence have been offered. Most of these definitions focus on intellectual achievements and deliberative thought, but as we have seen, intelligence requires more.

In chapter 3 we introduced Robert Sternberg's triarchic theory of intelligence. His definition, unlike many others, concerns how intelligence operates rather than just how it is measured.

Sternberg argues that intelligence consists of three types of adaptive capabilities: analytic, creative, and practical intelligence. Analytic intelligence is the kind that most of the other definitions emphasize, but the other two facets also play critical roles. Analytical intelligence focuses on abstract thinking, logical reasoning, and verbal and mathematical capabilities, the familiar components of intellectual achievement.

Practical intelligence includes tacit knowledge, which is often called common sense. Conversations and problem descriptions generally do not include this tacit knowledge because we assume that people will already know it. Tacit knowledge is usually acquired without formal training. Tacit knowledge is rarely discussed, in part because it is so difficult to articulate. How do you describe, for example, everything that you know about a peanut butter sandwich? As a result of this difficulty, the importance of tacit knowledge to problem-solving is often underestimated. But being difficult to describe does not mean that it is unimportant.

Sternberg also notes that intelligence is not just reactive but is also active in shaping its situations. Intelligent individuals do not simply respond to puzzles and problems; they actively seek them and seek to structure their environments to make it easier to address their issues. One way to solve a problem can be to change environments. Intelligence includes the ability to set and accomplish meaningful goals. Intelligent people can recognize the existence of a problem, define its nature, and represent it. They can recognize where knowledge is lacking and work to obtain that knowledge. Although intelligent people benefit from structured instructions, they are also capable of seeking out their own sources of information.

Sternberg's view of intelligence can be applied directly to artificial general intelligence. It helps to point out just where progress is needed most to achieve artificial general intelligence. Computers excel at analytic capabilities, so it is no surprise that computational intelligence shows strong success in this area. Analytic capabilities are where human and machine intelligence currently overlap the most. Current implementations come up short, however, in terms of practical and creative intelligence. These capabilities are still provided by human designers.

If all that is necessary for a machine learning system is to engage its analytic capabilities, then the machine is likely to exceed the capabilities of humans solving similar problems. Analytic problem solving is directly applicable to systems that gain their capabilities through optimization of a set of parameters. On the other hand, if the problem requires divergent thinking, commonsense knowledge, or creativity, then computers will continue to lag behind humans for some time. These latter properties are also required for general intelligence.

Another so far unsolved problem for the definition of artificial general intelligence is: just how general does general intelligence have to be? Some definitions suggest that a generally intelligent machine should be able to do any cognitive (as opposed, perhaps, to motor) task, to solve any problem, that a human could. That definition is perhaps so broad that a human might not be able to qualify as generally intelligent. In contrast, it seems that the higher a person's skills get in certain areas, the more narrowly those skills are focused. No one person can solve every problem or solve every problem equally well.

Perhaps by "general," we mean that our artificial general intelligence should be able to solve any kind of problem at some time, but not every kind of problem at the same time. Effective problem solving in humans seems often to take years of education or practice for the person to become an expert. No one can be an expert at everything. It remains to be seen just how general an artificial general intelligence needs to be.

We might like to say that to be general, the intelligence must be autonomous. If current versions of specific artificial intelligence get their intelligence from the structure given to them by humans, their general intelligence is that of the human designer, not the computer. Again, it is not clear how autonomous an artificial general intelligence would need to be.

Divergent thinking implies that the computer can address problems for which it has not been specifically designed. More critically, divergent thinking means that the computer can address problems using methods that have not previously been associated with that problem, or perhaps with any other problem. It can create new approaches to solving the problem. Computers are very good at convergent thinking, where they execute a series of steps to solve a problem, but they are not good at autonomously finding out what those steps should be.

Achieving General Intelligence

There are three perspectives on our prospects for achieving artificial general intelligence. According to one view, achieving general intelligence just takes more of what has proved successful for task-specific intelligence—a taller stack of hedgehogs. According to the second view, general intelligence cannot be accomplished by machines, because general intelligence requires human consciousness or some other ineffable quality that only humans can have. According to the third perspective, artificial general intelligence is possible; however, it requires some new developments that are not yet available but can be outlined. I fall into the third perspective.

Jürgen Schmidhuber's (2009) Gödel machine is an example of the stack-of-hedgehogs perspective. The basic idea is that general intelligence can occur in a system that consists of a collection of modules, each of which solves a specific kind of problem, and an overmodule that selects and coordinates these specific modules to allow the solution of problems that were not specifically addressed by one of the modules. This overmodule has exactly the same kind of structure as the task-specific modules, but its goal is to select and coordinate among the specific modules. It learns from its experience deploying the specific modules and the outcomes each one produces. In other words, general intelligence is specific intelligence applied to the problem of selecting specific problem-solving mechanisms.

With enough computational capacity and enough time, such a system could do anything. Optimization in this view is the lever that we could use to move the world. I will have more to say about Schmidhuber's ideas later in this chapter, when we consider what it will take to achieve artificial general intelligence.

From the ineffable-consciousness point of view, artificial general intelligence can never be accomplished because it requires some property that can only be found in humans. At best, machines may simulate having this property, but without it, they cannot be truly intelligent.

Hubert Dreyfus's work, discussed in chapter 9, is an example of this kind of approach. Another example is Roger Penrose and Stuart Hameroff's notion that consciousness somehow involves quantum mechanical coherence in the microtubules of the brain. Because Penrose is a renowned physicist who has worked on such weighty problems as black holes and knotty ones like string theory, he looks to the properties of quantum mechanics to explain consciousness. Both Dreyfus and Penrose imply that consciousness is critical to intelligence and there is something mysterious about consciousness that cannot be explained in terms of computational methods. John Searle, creator of the Chinese room thought experiment, also believes that there is something about brains that allow them to have symbols with a certain kind of aboutness (philosophers call it intentionality) that is critical. Computers, he argues, are purely syntactic, so they have no access to meaning. They can only follow rules relative to symbols, but the symbols have no meaning for a computer. Meaning is essential to human intelligence, and only brains can do it.

Both approaches—hedgehog stacking and ineffable consciousness—are wrong. The hedgehog approach is wrong because optimization is limited to adjusting parameter values, and the ineffable-consciousness approach is wrong because it does not really say anything useful about intelligence. It just says that computers can't do it.

The third approach to general intelligence supposes that artificial general intelligence does require some mechanisms that are not currently available, but it assumes that with some amount of research, these mechanisms could be developed. In the rest of this chapter I will try to sketch out what such a research program might look like and, in that context, extend the critique of the hedgehog perspective.

Beginning the Sketch of Artificial General Intelligence

In chapter 3, we began a discussion of just what general artificial intelligence would look like. Among the kinds of skills that an artificial general intelligence agent should have are:

- the ability to reason
- the ability to engage in strategic planning
- the ability to learn
- the ability to perceive
- the ability to infer
- the ability to represent knowledge

But even this partial list of skills does not distinguish between specialized computational intelligence and generalized computational intelligence. A chess-playing program, for example, could easily be said to have these properties but still be entirely specialized for chess playing.

To those skills, I would add abilities like these:

- the ability to learn from a small number of examples
- the ability to identify problems
- the ability to specify goals
- the ability to find new and productive ways to represent problems
- the ability to create new knowledge representations and structures
- the ability to compare multiple approaches to a problem and evaluate each one
- the ability to invent new approaches
- the ability to think about ill-formed, vague ideas and make them actionable
- the ability to transfer knowledge from one task to another
- the ability to extract overarching principles

- the ability to speculate
- the ability to reason counterfactually
- the ability to reason nonmonotonically
- the ability to exploit commonsense knowledge

Computers are much better at calculating than humans are. They are more systematic, more algorithmic than people are. They do not get distracted. The kinds of tasks where computers can surpass human abilities are those that are well-structured, are finite, and can be learned using optimization of a model's parameters.

There are other problems, however, that cannot be described in the same way. These are poorly structured problems, perhaps of unknown scope, for which we cannot readily measure progress toward a goal, or that cannot be well-specified for a number of reasons. At this point, for example, even a multilayered machine learning system that learns to select machine learning methods can only select among what it already knows; it cannot generate novel approaches. It may combine old parts in new ways, but genius often requires new parts.

A 2016 survey of Millennials (World Economic Forum, 2016) identified the problems that they thought were most significant. These problems include:

1. Climate change and destruction of natural resources (45%)
2. Large scale conflict and wars (38%)
3. Religious conflicts (34%)
4. Poverty (31%)
5. Government accountability, transparency, and corruption (22%)
6. Safety, security, and well-being (18%)
7. Lack of education (16%)
8. Lack of political freedom and political instability (16%)
9. Food and water security (15%)
10. Lack of economic opportunity and unemployment (14%)

These problems are ill-formed and underspecified. They have no definitive formulation or path by which they could be solved. It is difficult to measure whether an attempted solution to one of these problems is actually moving it toward solution, is making it harder to find a solution, or is doing nothing at all. There is no easy way to definitively determine even whether they have been solved.

Contrast these problems with playing games, like checkers, say. Current approaches to computational intelligence have been very successful at solving problems like games but are helpless when faced with problems like those on the Millennials' list.

Even problems like the one involving hobbits and orcs crossing a river are easy for a computer to solve if the problem has been represented properly, but so far, it takes a human to create that representation. In an important sense, the intelligence is in how the problem is represented. New computational approaches are needed to address problems like these.

Well-defined problems come with initial states, where we are now, and goal states, where we want to be. For example, we have a collection of photos, and we want to be able to identify which of those photos contains a cat. Or we have a chessboard, and we want to be able to win the game against the best opponents we can find. Even self-driving cars have clear evaluation methods, although these are more challenging. We can compare any two computer systems and decide which of them is superior.

Other functions that we would expect of an intelligent system are not so easy to assess. We could write a computer program that would generate paintings, but it is not obvious how to evaluate the success of that system. A computer program to paint in the style of van Gogh, for example, is relatively straightforward, but a program to create new paintings or new styles of painting is much more difficult to assess.

It is a serious mistake, however, to assume that our ability to assess a problem solution is related to the importance of that problem. It is equally an error to assume that the problems that are easy to evaluate are typical of the kinds of problems that a generally intelligent agent would need to solve. Computers need to deploy the kind of insight that converted chess playing to tree navigation to be generally intelligent.

Easy-to-evaluate games like chess, checkers, or go are formal, well-structured, and full-information problems. They are fully described by their rules and their current state. They can be treated as a purely mathematical process. They do not depend on any physical instantiation of a game board. One could play any of these games without seeing any physical game pieces. High-level chess players can play chess blindfolded, for example. On September 24, 2016, Grandmaster Timur Gareyev achieved a world record by playing 64 consecutive blindfold chess games, winning 54 of them.

It is the form of the rules, rather than the physical properties of the game pieces, that make up chess. A computer could continue to play chess if all of the rest of the universe disappeared. The computer could determine whether it had won or lost a game purely by maintaining an accurate representation of the game state.

Jonathan Schaeffer figured out that there were about a quadrillion unique checkers positions that had to be evaluated to prove that any given move was, in fact, the best one possible. After working on this problem for about 18 years, he was able to evaluate all of those moves and was able to prove the optimality of each choice in each state.

Evaluation of checker moves is slow because of the large number of combinations of future moves that are possible in a game of even that complexity. But it is still a formal, full-information game, and even a complete analysis does not break new ground in a search for artificial general intelligence. Most real-world problems cannot be reduced to formal problems. Even Bostrom's paper-clip collector problem cannot be reduced to a formal problem. No one could prove that the paper-clip collector was doing the best job possible.

An important exception to the widespread focus on formally structured full-information problems is self-driving cars (see chapter 6). Driving is not formal; the world absolutely does matter. At best, the problem of driving is semistructured. But it also suffers from another problem. The vehicle must not only solve computational problems but has to navigate in a dynamic physical world with sensors that can be inaccurate. Sensors are imperfect; unexpected things happen in the world. The state of the world, and not just the state of the computation, determines the success of driving.

The actions the vehicle takes are imperfectly related to the state of the world in which they occur. For these reasons, self-driving cars present a distinctly different computational intelligence problem than do chess-playing computers. Even if someone spent 18 years at it, there is no way to prove that any action taken by the vehicle is the best possible action.

Self-driving vehicles take time to train, and they take time to test. Tests of the success of the system are less well-defined than they are in chess, but it is still practical to test autonomous vehicles by letting them drive. Driving has measurable consequences. We can tell whether the vehicle collided with an obstacle, for example. These factors make this one of the most interesting computational challenges around today, but it still is not on the road to artificial general intelligence.

The gamelike problems discussed so far can be described as a search through some kind of solution space. In chess and checkers, the solution space is the tree of potential moves. In self-driving cars, it is the set of predictions of obstacles and other kinds of events. The insight that allowed self-driving cars to be successful is the representation derived by Sebastian Thrun and a few others that allowed the vehicle to take advantage of unreliable evidence about its surroundings and to use one set of sensors to provide critical feedback about the predictions made by another set. Self-driving cars depend on a number of machine learning applications, each of which solves a simpler more or less structured problem.

Other problems, like the parking question, require a different set of capabilities: *The downtown area of your city does not have enough parking spots available. What do you do to resolve this situation*? Although humans might someday be able to break a problem like this one down into combinations of more or less structured problems, each of which could be solved with machine learning, the effective representations for them remain unknown, as does the process for rerepresenting them as solvable subproblems. At best, solving them using computational intelligence will require the invention of new representations, presumably by inventive computer scientists.

Insight problems present a deeper challenge to computational intelligence. Examples like Maier's two-strings problem (see chapter 2) are among the kinds of problems that a general intelligence will need to be able to solve. Solving insight problems requires the solver to create an appropriate representation. Once the right representation is achieved, solving the problem is almost trivial. Much of what we generally think of as the highest human intellectual achievement depends critically on the person creating a new representation for a formerly resistant problem.

Friedrich August Kekulé reported, for example, that he came up with the idea of representing the structure of the organic chemical benzene as a ring as a result of dreaming about a snake swallowing its own tail. Russian chemist Dmitri Mendeleev said that he created the periodic table of elements also after a dream.

Mendeleev's original table, published in 1869, was arranged by atomic weight (approximately proportional to the number of protons and neutrons in an atom), but it had a few exceptions, where the properties of the elements suggested a reversal of the elements within a row. He also left gaps in the table, which suggested the existence and properties of elements that had not yet been discovered. His second table, published in 1871, arranged the elements by atomic number (the number of protons in the atom). This table is essentially the one we know today.

The important thing about Kekulé and Mendeleev's achievements is not that they occurred to their inventors as dreams but that the two men achieved a new and useful representation. Although Mendeleev worked a long time with different arrangements of the elements, he did not report finding any that he thought were closer to being correct until he achieved the 1869 arrangement. The suddenness suggests that his new representation was not achieved by any kind of optimization process, such as gradient descent. It was difficult to measure progress in finding an appropriate representation. But eventually he did come up with a representation that worked.

The suddenness, of course, of creating these representations is based on the scientists' self-reports. We do not know what "unconscious" processes occurred that led to the dreams or the creations. We know, for example, that Mendeleev had been working for a long time on the problem of how to arrange the elements by their chemical properties. We know that he thought that the card game "patience" was somehow suggestive of a solution, but none of that was fitting together to give him an answer that he thought was reasonable. Our ability to create computational approaches that were able to come up with new representations would benefit greatly if we could understand what those processes were.

All of these forms of problem solving depend on a human discovering or inventing a new representation. At present, this requirement is the source of all innovation in machine learning and its biggest bottleneck. It's fine to fantasize that computers will be able to create their own intelligence, redesign themselves, and do so at an ever increasing rate, but at this point and for the foreseeable future, that is just a fantasy or, according to some, a nightmare.

An artificial general intelligence will need to be able to create its own novel representations. This is one of the most important areas where effort is needed.

In the 1980s, and again in the 2000-teens, there was a show on US television in which the title character, MacGyver, would come up with unique ways to solve problems with whatever was at hand. That is a dramatic example of the kind of problem-solving that an artificial general intelligence will have to solve.

Another example of this kind of problem solving might be the creativity exercise uses for a brick. How many uses can you think of for a brick? There are common uses for a brick, but if you think about it a little, most people can come up with some unusual ones—how about for cooking a chicken? That is but one example of the kind of problems that an artificial general intelligence will need to solve.

More on the Stack of Hedgehogs

As we discussed earlier in this chapter, one suggested approach to creating an artificial general intelligence agent is to layer a high-level machine learning module to select and combine more specific modules. Under this view, general intelligence would require just the same kinds of processes that are required for specific forms of intelligence. This approach assumes that general intelligence can be solved by adding more parameters to the current kind of single-task systems in use today.

The stack-of-hedgehogs or more-of-the same approach rests on a few assumptions, as described by Cassio Pennachin and Ben Goertzel (2007). They define intelligence as the maximization of a certain quantity by a system that is interacting with a dynamically changing environment. They also note that this approach rests on the validity and applicability of the Church–Turing thesis (discussed earlier, in chapter 3).

The Church–Turing thesis can be summarized as the claim that anything that is computable can be computed by a system with a small set of simple operations. Alternatively, any computable function can be computed by a machine with the capabilities of a Turing machine. The word "computable," in this context, has a special technical meaning. A computable function is a predetermined step-by-step procedure that is certain to produce a verifiable answer in a finite number of steps. A computable function is an algorithm, for which a certain set of inputs, when the proper processes are executed, will result in a specific output. Another way of saying this is that every computable function is a form of logical deduction (Copeland & Shagrir, 2019).

The Church–Turing thesis is critical to computational intelligence because it makes clear that two computational systems with equivalent capabilities are equivalent machines, no matter how they are each constructed. As a result, if intelligence is a computable function and if it is computed by brains, then it is perfectly reasonable to expect that a system built on silicon should also be able to compute the same function if it has equivalent power. Therefore, if brains compute, a Turing machine must be able to execute the same function.

The key assumptions here are:

1. Intelligence is a function that can be implemented by an algorithm. Intelligence is computable by a Turing machine.
2. The intelligence function takes as its input an instance of the problem and returns a solution.
3. The intelligence function is a process of optimization—maximizing of a certain value by a system interacting with a dynamic environment.
4. A Turing machine can verify the correctness of this solution.
5. The brain computes with no more capability than a Turing machine. With enough computational capacity and enough memory, some implementation of that machine will be able to compute the same function as the brain. Some computer equivalent to a Turing machine will be able to compute the function computed by the brain.

The first assumption of this approach is reminiscent of John McCarthy's original hope (in the proposal for the 1956 Dartmouth workshop) that it would be possible to describe the features of intelligence in sufficient detail to be able to get a machine to simulate it. I would argue, however, that it is a fundamental error to assume that intelligence is an algorithm—a specific sequence of steps that always returns the correct answer. This assumption asserts that intelligence would have to be a form of mathematical deduction, whereas machine learning is a process of induction. On the basis of training examples, the computer predicts how subsequent unseen items will be classified, for example.

It's important to be clear here about just what I am claiming. Computers and brains both use algorithms, but intelligence is not itself an algorithmic process in that it cannot be considered infallible.

To quote Alan Turing (1947):

> It might be argued that there is a fundamental contradiction in the idea of a machine with intelligence. . . . It has for instance been shown that with certain

> logical systems there can be no machine which will distinguish provable formulae of system from unprovable, i.e., that there is no test that the machine can apply which will divide the propositions with certainty into these two classes. Thus if a machine is made for this purpose it must in some cases fail to give an answer. On the other hand if a mathematician is confronted with such a problem he would search around and find new methods of proof, so that he ought eventually to be able to reach a decision about any given formula. . . . Instead of it [the machine] sometimes giving no answer we could arrange that it gives occasional wrong answers. But the human mathematician would likewise make blunders when trying out new techniques. . . . In other words then, if a machine is expected to be infallible, it cannot also be intelligent. There are several mathematical theorems which say almost exactly that.

The mathematical theorems that Turing is referring to are probably Gödel's incompleteness theorem and Church's and Turing's theorems that some problems are undecidable by a Turing machine. For any sufficiently powerful formal logical system, there will be statements that are true but cannot be proven to be true by the system. There are fundamental limits to formal systems that cannot be overcome within the context of formal systems. So, while formal systems, such as logic, are important to intelligence, they are not enough for intelligence. They are incomplete.

The second assumption is also faulty if the goal is to create artificial general intelligence. The second assumption asserts that the intelligent agent is given an instance of the problem it is to solve. That is fine for solving individual selected problems, but a generally intelligent agent should not have to be handed a structured representation of a problem. It should be able to find its own structure.

The third assumption is where more-of-the-same gets its name. The same process that allows machine learning to solve specific problems is asserted to be sufficient to solve general intelligence. It is the assumption that the system already has all of the tools it could need to solve any problem.

According to the fourth assumption, some algorithm should be able to verify the correctness of an intelligent solution to a problem, but intelligent solutions to real problems other than puzzles and games and the like are often difficult to assess and impossible to verify. Intelligence often involves predictions that require estimations and are inherently uncertain.

Of these, only the fifth assumption is more or less reasonable. Turing machines can compute universally any computable function, but that does not mean that they are then incapable of operations that are not strictly algorithmic. As current machine learning demonstrates, computers are capable of inductive inference (see chapters 4, 5, and 6), not just deductions. They can infer from examples to rules, not just apply rules to make decisions about examples.

As it happens, these assumptions are fine for achieving the kind of specialized problem-solving that has been so successful over the last several years. Where they mainly fall short is in the unstated sixth assumption that these are all that is needed for general intelligence. The Church–Turing thesis, I argue, has misled computer scientists into thinking that general intelligence can be deduced from specialized intelligence, but general intelligence needs more. We take up that topic next.

General Intelligence Is Not Algorithmic Optimization

The Church–Turing thesis conceives of intelligence much too narrowly. Intelligence, like that shown by Einstein, but also like that shown by people every day, does not derive only from following a well-trodden path of specific instructions or muddling through the selection of one of several paths. General intelligence consists precisely of finding a new set of instructions, if you will.

Even small variations from the exact problem on which current computational intelligence programs have been trained can completely baffle them. Self-driving cars, for example, have been "deceived" into thinking that stop signs are really speed limit signs just by putting a small stickers on the sign.

Optimization is the process that modifies the values of parameters in order to maximize or minimize some value, such as error. Specific problem solving can be successfully addressed by such mechanisms if the optimization process is given an appropriate set of parameters to work with. Optimization does not create parameters; it works to adjust the model parameters it is given by the program designers.

Intelligence and TRICS

The hedgehog approach, as described above, assumes that the function that is implemented by a general intelligence agent takes as its input an instance of a problem and produces a solution as its output. More properly, we should say that the computer takes as its input a representation of the problem and produces a representation of the output. Computers cannot deal directly with baseballs, hobbits, or strings. Rather, these objects have to be represented in some mathematical form. Problems too have to be represented in some mathematical form. Chess could be represented as a mathematical version of psychological war between adversaries, or it could be represented as a tree of potential moves. Categorizing cats versus dogs represents the photographs as an array of numbers and represents the neural network as another set of numbers. Optimization sets the values of the numbers representing or implementing the network, but it does not affect the kind of numbers they are or the kind of raw representation that the problem starts with.

How the objects and the problem are represented is key to finding a solution to the problem. These are the representations it critically supposes (TRICS; see chapter 9). The solution is constrained by and contained in these representations. If the problem has to be prerepresented for a hedgehog in order for the system to address it, then the general intelligence comes from the designer of the problem, not the system. Without the ability to construct its own representation of the problem, it is literally game over for a more-of-the-same system.

From the hedgehog perspective, general intelligence would emerge from a layered system where a high-level module solves the problem of selecting special-purpose modules. The input would be a specification of the problem, and the output of this supervisory module would be the special-purpose module that solves it.

Each of the special-purpose submodules must have a well-formed solution space, and the overarching module for selecting or combining them must also be well-formed. The supervisory module is limited to selecting or perhaps combining those tools that it has in its collection of modules. It is dependent, then, on having a complete set of modules, because it cannot entertain the "thought" of creating a brand new module, it is incapable of inventing its own. It is limited to selection, according to the more-of-the-same framework, and selection is not generation.

One example of such a hierarchical system is Schmidhuber's (1996, 2009) so-called Gödel machine. In his view, this machine is a "class of mathematically rigorous, general, fully self-referential, self-improving, optimally efficient problem solvers.

Inspired by Kurt Gödel's celebrated self-referential formulas (1931), such a problem solver rewrites any part of its own code as soon as it has found a proof that the rewrite provides improved lifetime future value. This Gödel machine is a thought experiment and has never been built, and it probably never can be built.

Schmidhuber supposes that the machine can learn new modules by rewriting its own code if it can prove that the rewrite makes it better able to achieve its lifetime goal. As mentioned earlier, a system's programming, its code, is typically far less important to any computational intelligence system than are the data used to train it.

In any case, the Gödel machine suffers from at least four fatal flaws: (1) the need for proof before it modifies its code, (2) its reliance on lifetime future value to decide whether to rewrite its code, (3) the idea that it can be designed with a complete enough set of modules, and (4) the combinatoric requirement to evaluate many alternative modules and module combinations.

It's not clear how the machine will know which changes to assess for their future value. It is chooses changes randomly, it could take a very long time to find one that is successful. Even if a change does appear to be successful, actually proving that it is successful is impossible because it depends on a measure of lifetime future value. The only way that one can obtain an actual measure of lifetime future value, though, is to actually follow through to the end of the lifetime, but by then the opportunity for making any change is long past. So, the Gödel machine must estimate future value, but estimates are fallible and cannot be proofs, because they are always inductions, not deductions. That seems to be a contradiction. If making a change depends on proof, but proof cannot be obtained, then the system could never make any changes.

The biggest problem with the proposed Gödel machine, however, is the idea that we can provide it with a sufficiently complete set of elementary problem-solving techniques that it could actually use to solve unanticipated problems. It is ironic that Schmidhuber would call his machine a Gödel machine when Gödel's incompleteness theorems prove that such systems cannot be complete. No formal system, and the Gödel machine is a formal system, can be complete, and no formal system can prove its own consistency. We have already seen the impossibility of listing out a complete set of commonsense facts (see chapter 7). The idea that there is some set of elementary problem-solving methods from which all other problem solving could be derived is equally unlikely.

Schmidhuber's machine is designed to be purely deductive in Copeland and Shagrir's sense. If we had a set of elementary problem-solving methods, it is not clear how they could be combined deductively to actually solve problems. Rather, the machine would have to create hypotheses that might or might not be correct and then evaluate those hypotheses against the actual problem situation. It could, perhaps, combine modules in novel ways, because it is limited to just those modules it has. It is limited to finding a path through a specific parameter space. It still cannot create new spaces, but that is just what is required for intelligence.

A hierarchical system, like the Gödel machine, would suffer from two additional problems: the time it takes to discover whether a module could actually solve the problem and the combinatorics of trying different groups of modules to attempt to solve the problem. Without some powerful module-selection heuristics, which are incompatible with the deductive structure of the Gödel machine (because the heuristics cannot be proved correct), such a hierarchical system would find itself lost in thought. The heuristics used in current narrow problem solving are the result of human analysis of the properties of the problem being solved. These heuristics are built into the system's representation of the problem. It is not clear where such heuristics would come from in the case of a hierarchical system.

Learning which modules to apply in any situation would necessarily require a great deal of failure. When a module navigates its problem space, a single step in the wrong direction might not be very costly. Supervising thousands or millions of modules would take considerably longer. Each module would have to be trained on perhaps thousands or even millions of training events. If the wrong model is chosen, it may take millions of training examples to discover that it is, in fact, wrong. The nature of problems suggests that these alternative solutions would often have to be run sequentially, and so the system would have to wait until one is done before trying another.

Such a volume of effort would challenge even the largest computer networks and would inevitably take considerable time. A brute-force optimization plan simply would not be viable for a hierarchical module system. More complete systems might be able to solve more general problems but take eons to solve them. Recall the sum of three cubes problem discussed in chapter 11. It took 23 processor years to solve a very simple problem with only three variables. Problems with thousands of potential solutions would take an indescribable amount of time to evaluate by brute force. No matter how many hedgehogs we stacked up, we still would not be able to achieve a fox.

Transfer Learning

Any system attempting to implement artificial general intelligence will need to learn from its experience. But learning to solve one problem may interfere with solving others.

Google's DeepMind team has been using reinforcement learning to train a network to play vintage video games. In one experiment, they trained the system to learn 49 video games in succession, but each time the system learned a new game, it "forgot" how to perform with the previous one. It started from scratch every time it learned a new game. The problem of losing previously learned tasks when learning new ones is called catastrophic forgetting.

Rachit Dubey and his colleagues studied how long a reinforcement learning system took to learn video games like those studied by the DeepMind team. In one experiment, people took about 3,000 action units to learn to play a game, but the computer took about 4 million actions (an action was a key press, for example). When Dubey and his colleagues changed the appearance of the game elements, things took an interesting turn.

Their game contained primitive low-resolution graphics, but with recognizable objects, such as ladders, keys, spikes, and doors. When the experimenters modified the appearance of these objects, so that they could not be immediately recognized by a human player, play became much harder for the humans, but not for the computer. Depending on the exact manipulation, the human time to learn the game increased to 20 minutes, whereas the machine learning time remained approximately constant for most manipulations. People took advantage of their commonsense knowledge that doors open, that ladders could be used for climbing, but the computer did not have this background knowledge and so was unaffected by the manipulation.

Then another new game was shown to users in which getting to the princess was one solution, but other solutions were possible. The people who learned the game focused on getting to the princess and failed to even explore hidden reward locations. A randomly started machine, in contrast, tended to find these additional rewards because it did not have an expectation that the princess was the goal of the game. People could transfer their knowledge of other games that they had played and their knowledge of how objects of the world can be used to support actions, but this knowledge was not always a benefit.

The phenomenon of negative transfer is a well-known issue in human problem solving. The early Gestalt psychologists studied what came to be called the water jar problem, originally described by Abraham Luchins in 1942. In this problem there are three jars, each of which holds a certain amount of water, and the goal is to end with one of the jars holding a specific amount.

One of Luchins's problems involves a 29-liter, a 3-liter, and a 21-liter jar. The goal is to end with a jar containing exactly 20 liters. It would be difficult to measure out exactly 20 liters with only a single jar, but by pouring water from one jar to another, the problem can be solved.

Think for a moment about how you would solve this problem. The water jar problem is well structured and has perfect information. All of the information you need to solve the problem is contained in the description. It is a formal problem in that you can solve it without actually having to deal with jars or water. It depends on the properties of arithmetic, not on the properties of water.

Here is how to solve this problem (for clarity, let's label the jars A, B, and C, respectively):

Fill the 29-liter jar, A.

Dump water from jar A into the 3-liter jar, B, leaving 26 liters in the first jar.

Empty the water from jar B.

Dump another 3 liters from jar A into jar B, leaving 23 liters in jar A.

Empty jar B.

Dump another 3 liters from jar A into jar B, leaving 20 liters in jar A.

Problem solved.

Table 6 shows a series of 10 problems. Take a moment and solve these problems, if you will.

People often get faster at solving these problems as they move through the list. They show positive transfer from one problem to the next. A computer using reinforcement learning would take about the same amount of time to solve each problem. It is not clear how machine learning, as

Table 6

Problem	Capacity of Jar A	Capacity of Jar B	Capacity of Jar C	Goal Quantity
1	21	127	3	100
2	14	163	25	99
3	18	43	10	5
4	9	42	6	21
5	20	59	4	31
6	23	49	3	20
7	15	39	3	18
8	18	48	4	22
9	14	36	8	6
10	28	76	3	25

currently employed, could be used to provide positive transfer from one problem to the next without some explicit design for this specific set of problems. The people who come to participate in such a study do not usually have any specific training or knowledge of these problems before they start, except perhaps if they are serious fans of the *Die Hard* movies, where a problem like these is part of the plot of *Die Hard with a Vengeance*.

Problems 1–9 can all be solved using the same set of moves: Fill jar B, subtract jar C twice, and subtract jar A once (B – 2C – A). For problems 1–5, this pattern is the simplest way to solve the problem. Problems 6–9 can be solved using this same set of moves, but they can also be solved using a much simpler pattern of moves. Problems 6 and 9 can be solved using the moves A – C. Problems 7 and 8 can be solved using the move A + C. Because problems 6–9 can be solved using the same pattern as the earlier problems, participants rarely recognize that there is, in fact, a simpler way to solve the problem; 83% of participants used the same set of moves, (B – 2C – A), on problems 6 and 7, and 79% used it on problems 8 and 9. Perhaps surprisingly, a full 64% of participants did not solve problem 10 at all. People who got only problem 10, on the other hand, were overwhelmingly (95%) able to solve it, but the participants who were given the first nine problems were not. Luchins called this failure to solve problem 10 the Einstellung effect, or functional fixity. People performed the metatask of generalizing from one problem to the next; they had no reason to challenge their generalization on problems 6–9, and they tried to apply the same method to problem 10 where they failed.

This set of problems illustrates that transfer learning is not necessarily as straightforward as one might hope. It can be useful to solve problems, but it can also interfere with solving them. These problems show that people can be very good at transferring useful information from one problem to the next, at least under certain circumstances. The more similar the two problems are in their surface structure (for example, they all involved jars and water), the more likely the people are to be able to identify the analogy. We enhanced the similarity when we labeled the jars A, B, and C, which made the analogy from one problem to the next more obvious.

But these problems also demonstrate a phenomenon called confirmation bias. People tend to look for information that confirms their beliefs rather than information that challenges them. Problems 6–9 were consistent with the beliefs that they had extracted from the first five problems, so there was little reason to lead the solvers to find a simpler solution.

The right analogy can be helpful, but the wrong analogy, as for problem 10, can be harmful. When Luchins told the participants after problem 5 "Don't be blind," a full half of them found the simpler solutions for the remaining problems.

Confirmation bias is another heuristic that may be useful to select an effective module, but it can be a problem when it prevents the system from considering methods that would be obvious without the bias. Bias helps to solve problems, except when it does not.

A current machine learning system might be designed to address this kind of transfer of learning situation for this suite of problems. Each water jar problem has a clear state space and clear methods for moving from one state to the next. Reinforcement learning would probably suffice as a training mechanism. But designing a system that is not specific to these particular problems or even this kind of problem remains a challenge.

Transfer from one problem to the next depends on the similarity of the two problems, but similarity is itself a difficult concept. In principle, the more features two items share, the more similar they are, but as we have noted, any pair of items shares an infinite number of features. People seem to select a subset of them for any comparison. In current machine learning projects, the features to be compared are selected by the designer. An artificial general intelligence agent would not, presumably, have the benefit of a designer for unanticipated situations and so would have to find its own way to select relevant features.

Consider the problem where three cannibals and three missionaries arrive at a riverbank. They want to cross the river, and there is a boat that can hold two people. If at any time on either bank, the cannibals outnumber the missionaries, the cannibals will kill and eat the missionaries. How do they get across the river?

If you remember how to solve the hobbits and orcs problem (see chapter 2), then this problem will seem similar and easy. In both cases, we make some commonsense judgments that the river could not be crossed without the boat, and so on (see chapter 7). These assumptions are not stated in the problem description, and it is not clear how a machine would know them.

Another assumption is that hobbits and orcs are immutable. A hobbit cannot become an orc, and an orc cannot become a hobbit. But that assumption is not valid for the cannibals and missionaries. The missionaries are presumably in the land of the cannibals to convert the cannibals—more commonsense knowledge. One way to solve the river-crossing problem would be to convert the cannibals so they would no longer be dangerous. Then it would be easy to get everyone across the river safely. The similarity of the two problems could then get in the way of solving the hobbits and orcs problem if one first learned the conversion solution to the missionaries and cannibals problem. Because orcs cannot be converted, the problem solver might be stymied by attempting to transfer what was learned about missionaries and cannibals to the hobbits and orcs problem.

Expert problem solvers address the transfer problem by using more abstract knowledge of the problem. Relative to novices, they are less affected by the surface properties of the problem and more affected by the physical principles that they entail. Artificial general intelligence will require some abstraction methods to be able to improve the quality of the transfer learning that they employ. They will probably require some theory of the domain in which they are solving problems, a theory gained through experience. A theory is a representation that is more principled and more abstract than a catalog of observations.

Intelligence Entails Risk

Einstein did not come up with the photoelectric effect or his theory of general relativity by searching a space of potential parameter values. A scientific theory is a new way of representing the part of the world with which it is concerned. And because scientific theory building is one of the highest recognized forms of intellectual activity, it is useful to consider what we know about constructing such representations. We can leverage the analysis that was applied to understanding how scientific theories are created to help us understand how intelligence more broadly might be constructed.

For example, about the time that Einstein was coming up with his greatest work, a group of philosophers, the logical positivists, were working toward a goal of making scientific theories more consistent and logical (see chapter 2). The theories of relativity and quantum mechanics disrupted the core of physics as it was then understood. The positivists assumed that there must have been something wrong with the practice of science that let physicists deceive themselves into believing that the Newtonian view was correct. These philosophers set about developing an approach that would prevent them from ever being deceived again like that.

The logical positivists sought to remake science into a purely deductive process, like Schmidhuber's Gödel machine. They wanted to limit scientific statements to observation statements, like "That ball is red" and deductions from those observations. Their approach failed in part because there are no pure observation statements. Stephen Jay Gould, for example, discussed faulty observations in science that were used to support racial theories about intelligence. Other scientists subsequently criticized Gould's analysis (Lewis et al., 2011).

More critically, scientific theories depend on making predictions about things that have not yet been observed. Theories transcend observations and deductions. They reflect risky (meaning they could be wrong) inferences that are different from observations and deductions. They are extrapolations from models; they are not deductions from known observations.

Although there were observations that were consistent with Einstein's theories (for example, the Michelson–Morley experiment of 1887 on the speed of light), his theories were important not so much in describing what had been observed but in predicting what would be observed under specific conditions. Some of these predictions were not evaluated until 2016, about a hundred years after Einstein first proposed his theories of relativity.

He could have been wrong about his predictions, so they were not deduced from the observations already available; they were risky predictions that turned out to be correct. The theories were created, not deduced.

Predictions are necessary to intelligence, and deductions are not sufficient to yield them. How exactly a theory can be represented for machine learning and artificial general intelligence remains an open question.

Creativity in General Intelligence

Mozart is well-known today not for his ability to play the violin but for the music that he composed. Einstein won the Nobel Prize for his creative work on the photoelectric effect, which, one might argue, was one of his lesser creative accomplishments. Gerald Edelman won his Nobel Prize in Physiology or Medicine for his work on the immune system and its ability to learn. In fact, every one of the Nobel Prizes in science was awarded to someone who created an elegant understanding of a complex phenomenon. In artificial intelligence terms, every one of them created a novel and effective way of representing their problems.

This aspect of genius is currently missing from our computational intelligence systems. The fact that people can do it suggests that it is, in principle, possible for machines, but even for people, it does not come along frequently.

Good ideas—for example, new scientific theories or great musical compositions—do not come along every day. Many problems persist for years before someone comes up with a solution for them. While good ideas are not common, neither are they so rare as would be expected if they occurred by chance. The finding that important theories are often invented independently, but nearly simultaneously (I'm thinking of Charles Darwin and Alfred Russel Wallace both coming up with the theory of evolution at essentially the same time) suggests that something in the "air" encouraged both men to think along the same lines. Inventions and discoveries like this are clearly nonrandom, but neither do they occur on demand. As Pasteur said, "Chance favors the prepared mind," but we have yet to learn exactly what that preparation consists of or how to provide it to machines. And, is it really chance?

In some special sense, limited creativity by an artificial agent is very common. AlphaGo could be said to be creative when it made a move that, apparently, had never been made before by a human go player. The move was so shocking to its human opponent that he got up from the table and walked around for some time contemplating what it meant.

Celebrated examples of human creativity, such as a Mozart symphony or Einstein's theory, seem to be of a different sort than the surprising move that AlphaGo produced. A Mozart symphony is not just a deduction from past compositions, nor is it a simple extrapolation or recombination of what had gone before. Pablo Picasso and Georges Braque's creation of cubism, as another example, was a shocking departure from the approaches to art that preceded it. Really celebrated acts of creativity go outside the space of existing parameters. They create new sets of parameters.

When seen this way, creativity is not magic. It does not rely on any kind of miracle but on recombination and, more importantly, on reconceptualization. The trick in getting computers to be more than trivially creative is to figure out how they could be programmed to reconceptualize and change the rules or the space of a problem.

Growing General Intelligence

Education plays a crucial role in producing general intelligence in humans. Even though machine systems and human brains are very different, maybe if we gave machines the kind of educational experience that we give people, they would learn to be generally intelligent. In the context of human expert performance, for example, there is a lot of evidence that a certain amount of experience is necessary to achieve expert level performance.

In 1931, Winthrop Kellogg set about raising a young chimpanzee, named Gua, along with his young son, Donald. Kellogg was interested in the nature/nurture question of whether an ape raised as a human would come to act like a human. To Kellogg's disappointment, Gua learned many things faster than Donald did but never showed any interest in communicating in a humanlike way.

Alan Turing (1948, 1950) proposed a similar approach to creating artificial intelligence with a computer. Rather than try to build a fully adult intelligence in a computer, he talked about building a machine that simulated a child. With an appropriate course of education, he argued, it could grow into an adult intelligence.

Actually, he argued for creating a number of these child machines and comparing them one against another to identify the best methods to use. He saw this competitive process as similar to evolution but hoped that with direction it could evolve intelligence more quickly than evolution did.

If we focus on intellectual functions, then it looks like a child's brain is simpler than that of an adult. But Moravec's irony is that computer simulation or emulation of higher cognitive function is actually easier than simulating the kind of activities that children engage in. Processes like face or voice recognition, bipedal balance, and others that we usually ignore in artificial intelligence research are actually more difficult than playing chess or answering questions. We have begun to make progress on these, but that progress is relatively recent compared to the functions that we usually hold up as examples of intelligence.

Still, the idea of starting with a simpler system and letting its experience train it is a valuable idea, whether we faithfully simulate a child or a young chimpanzee. Machine learning can be a powerful tool in evolving an intelligence.

Nick Bostrom argued that to be effective, such a system would improve mainly through trial and error (which takes time) but would necessarily be "able to understand its own workings sufficiently to engineer new algorithms and computational structures to bootstrap its cognitive performance" (Bostrom, 2014, p. 29). It must be able to recursively improve itself.

Recursive self-improvement means for a system to update its state within its problem space. That is the definition of machine learning. It's not clear, however, what "understanding its own workings" might mean in the context of machine learning. Presumably, a system would employ one subsystem to evaluate its operational capabilities, identify its limitations, and work to overcome them. Overcoming them would presumably mean finding new representations and optimizations, instantiating new algorithms and cognitive structures—representations. Such capabilities are not even being investigated at the present time. It is easy to imagine how a metalearning machine that learned to improve itself might revolutionize artificial intelligence if it were to exist. But it is not at all obvious how one could actually be built.

Whole Brain Emulation

Arguably, the best model for general intelligence is the human brain. One approach to building artificial general intelligence is to emulate the human brain (see chapter 5). The idea is to duplicate as closely as possible the operations of each individual neuron and its connections. To the extent that we can emulate the entire brain, we should then be able to duplicate its function. The argument is that we would not really need to understand how the brain does its computations; rather, by building a machine that implements the same function at the level of neurons, we would automagically build a system that would implement the same intelligence.

Using the brain as a metaphor for computational intelligence has in fact shown itself to be a powerful tool. The neural network modeling that has been in widespread use since the 1980s has addressed many problems that were previously resistant to solution. But this level of modeling is very far from brain emulation. The neurons that are simulated and the structure in which they are organized are both distant approximations of how brains actually work. It is more correct to say that current neural network models are inspired by real neurons than to say even that they simulate them. Whole brain emulation, on the other hand, implies much more faithful reproduction of the structure and function of the brain than has been available in computational neural networks.

I am not at all confident that we will have a sufficient understanding of the human brain, its structure, and the functions of the neurons it contains any time soon. Neuroscience has made enormous strides over the last few decades, but in comparison to what we would need to know to emulate the brain, that science is still in its infancy, I believe. We don't even know how neurons store memories (Sardi, Vardi, Sheinin, Goldental, & Kanter, 2017). In chapter 5, we discussed an experiment that found that the memory stored in a neuron can change over time, even reversing (Driscoll, Pettit, Minderer, Chettih, & Harvey, 2017). We have had the full connection pattern of the neurons of the roundworm *C. elegans*, but we still cannot simulate its behavior.

We might soon have the computational capacity to emulate the human brain, but we are extremely far, I believe, from knowing what it is that we want to emulate. Understanding the operation of the brain continues to be helped by broadening computational capabilities, but those capabilities cannot solve the fundamental neuroscience issues that block our understanding of the brain. At this point, it is pure science fiction to think that we can emulate a brain, let alone record the state of the brain sufficiently to extract the personality from it and implement the personality in a computer as some people have suggested.

Analogy

Analogical reasoning is likely to be a key feature of artificial general intelligence. Until computers can solve analogy or abstract classification problems in a generic way, they will be limited to the navigating a predetermined space. Kekulé's dream led him to the structure of benzene because he saw a relationship between the snake eating its own tail and the structure of the benzene molecule. Metaphor use often entails finding a property that is shared between two things. In Kekulé's case it was the shape of the tail-eating snake and the shape induced by the physical forces that connected the atoms of the benzene molecule. Mendeleev saw a helpful analogy between a familiar card game and the arrangement of elements in the periodic table. The more surprising metaphors are those that involve an atypical feature that is common between two things. Surprising metaphors are the ones that are useful for creative thinking because they lead to unusual, and sometimes useful, ways to think about things.

A related potential source of ideas is jokes, particularly puns. There are several theories of what makes a joke funny. In the case of puns, the most likely one is incongruity theory. According to this hypothesis, a pun is humorous when it leads you think one thing and then discover that the word was used in an incongruous way. Here's one: "My ex-girlfriend misses me . . . but her aim is getting better." The setup of the pun leads the listener down a garden path, which has to be reanalyzed when the punchline reveals the incongruity.

The point of thinking about puns is that they indicate a way of thinking that has not been investigated in the context of machine intelligence. Puns expose a kind of ambiguity. Reducing that ambiguity when we hear the punchline highlights a different set of relations than we had in mind when we heard the setup of the pun. The new relations revealed by the incongruity can be the source of creative ideas. These new relations can lead to a reformulation of the problem's representation.

Similar kinds of reformulation could be useful for computational intelligence. The analogy between the dance party and the mutilated checkerboard in chapter 2 helped people solve the checkerboard problem. Analogy in the water jars problem both helped and hindered solutions of related problems. Can we find a mechanism that would allow machines to identify the right properties to include in an analogy and then take advantage of them to solve new problems? Simply supplying a computer with a set of analogies that it could apply to problems runs into the same incompleteness issues as other attempts to find an exhaustive set of primitives. As in other areas, it is unlikely that anyone could come up with some set of sensible primitives.

Finding analogies between problems is difficult for humans to do, and it remains a formidable challenge for computers. Current approaches to finding analogies include expensive hand-created databases. There are no known methods of using machine learning effectively to identify useful previously unknown analogies. But there is no principled reason why a computer could not eventually have this skill.

Other Limitations of the Current Paradigm

Machine learning systems are dynamic. Machine learning uses its optimization method to adjust the system's state, typically by small steps at a time, to better approach the system's goals. Machine learning is possible because of certain inherent constraints on what can be learned. For example, a machine that is learning a concept typically requires examples of the category members and examples of things outside the category. The members of each category are not completely arbitrary but are similar in some way to each other. The success of machine learning depends on this similarity assumption because it will classify unseen items by how similar they are to the learned categories. Without the assumption that similar items should be treated similarly, the best a machine could do would be to memorize the examples, and then it would fail completely to apply this knowledge to examples it had not seen.

Machine learning optimization also typically breaks the learning process down into small steps. At each adjustment, the system makes small changes to its state or parameters. Large changes have the potential to adjust the learning system from one poor state to another when there is actually a position in between the two that would be a better choice.

Machine learning, thus, depends on a "continuity" assumption. It assumes that similar items are to be treated similarly and that small changes will have small effects on its assessment.

A dynamic system is one where the state of the system changes over time as the result of interactions among its elements. For example, the number of bass fish in a lake depends on the rate at which parent fish spawn, the rate at which the eggs hatch, and the rate at which fish die. The population is described, then, as a dynamic system where hatching, spawning, and dying all depend on one another.

Machine learning systems are dynamic systems, but not all dynamic systems meet these continuity assumptions. Chaos theory (sometimes called the "butterfly effect"), for example, describes dynamic systems where small variations can produce very big changes. The behavior of these systems is easy to predict over very short time ranges but impossible to predict over longer time ranges. Even very simple systems can behave chaotically.

Chaos theory describes dynamic systems where rules determine how the system transitions from one step to the next. Chaotic systems are, therefore, easy to predict over the short term, because each step is governed by a specific rule. Over the long term, however, chaotic systems appear to be random because these systems are sensitive to very small changes and inaccuracies, such as rounding error. Chaotic systems are sometimes called the butterfly effect, because, in principle, the tiny effects of a butterfly flapping its wings in Brazil could ultimately affect whether there will be a hurricane in Florida. Small changes can lead to big seemingly random effects over time.

Edward Lorenz, one of the early pioneers in the study of chaos, first noticed it when doing a weather simulation in 1961. He wanted to replay part of a computerized weather simulation and restarted the simulation partway through by manually entering the numbers that had been printed out at that stage. To his surprise, the machine began to predict weather substantially different from that predicted in the earlier run. The reason for the difference eventually came down to the number of digits that he had entered when he restarted the simulation. The computer worked with six-digit precision, like 0.143234, but when the program saved the numbers from the previous run, it only printed out three digits, like 0.143. That difference was tiny, but in the context of weather and its inherent chaos, even such small differences led to huge variations in the predicted weather.

Chaos theory is important in the context of intelligence because it is an example of a situation that is so very different from the kind of behavior we observe in games and many other artificial intelligence situations. It violates the continuity assumption over substantial time frames, even as it adheres to it over brief time frames.

Chaotic behavior is common to many natural dynamic systems, including those that would be likely to concern an artificial general intelligence agent, including weather, road traffic, anthropology, sociology, population ecology, environmental science, computer science, and meteorology. Unlike games, life is more often characterized by dynamic systems, involving feedback loops and often involving chaotic behavior patterns. Artificial general intelligence will have to deal with these real-world phenomena, not just the well-structured patterns of games.

My larger point is that the kind of processes that govern games are not applicable to many of the other phenomena in the world. It is not just that other problems are more complex than games; it is that the kinds of processes that are effective for solving games are not likely to be effective for phenomena that involve hidden and uncertain information. Business negotiations, weather forecasting, elections, war, even neural activation may be more correctly described as chaotic. We will need different kinds of computational tools to address these situations, not just more of the same ones that let a computer win at go.

General intelligence will require a paradigm that enables the system to learn overarching principles from its experience on specific problems. That computer will have to understand novel metaphors and analogies. It will have to create its own problem representations. Current approaches to computational science begin with much of the problem already prestructured in the representation provided by their designers. An agent cannot be said to be generally intelligent unless it can structure its own problems.

A generally intelligent agent may not have to perform in the same way that humans do, but there is still much to be learned from human performance. A human child learns to identify rabbits after one or a few exposures. Deep learning systems may require millions of exposures. We will have to identify the constraints that let a human child jump to a conclusion without all of the painstaking effort involved in learning the right parameter values in the computer's deep learning network.

When computational systems do try to exploit the knowledge of humans performing similar tasks, they often depend on people to describe how they think they are accomplishing the task. But these reports are limited and unreliable. They are often rational reconstructions or convenient fictions describing what must have happened rather than describing what did happen. Clever experimentation often reveals that how people say that they do tasks does not correspond to objective measures of their performance. Many tasks cannot be described at all.

Metalearning

Metalearning is learning about learning or learning how to learn. Metalearning might be available to extend the kind of problems a computer might be able to solve, but metalearning is not without its own problems. Metalearning can interfere with problem-solving as well as support it. Recall the water jars problem. When transfer prevents people from finding an adequate solution, it can lead to artificial stupidity instead of artificial intelligence.

The standard computational approach to solving the water jars problem, though, would treat each problem independently. An engineer could design a method that keeps track of the effective moves from one problem and prioritize them in exploration of the state space. Such a system would solve the first nine of Luchins's problems using the same pattern. Because the pattern learned during the first five problems continues to solve the next four, this system, too, would fail to find the simpler solutions for the latter problems. It would initially fail on the tenth problem, but then it would eventually be able to find a solution because it only prioritizes, it does not eliminate, previously effective moves. After it solved problem 10, would it have forgotten how to do the earlier problems and need to start from scratch?

This engineering approach would be entirely specific to this particular set of problems. The current machine learning paradigm does not have a mechanism for abstraction. It would have to be explicitly designed as part of the problem representation, but an artificial general intelligence would have to be equipped with such a mechanism.

Without a common representation for the successive problems in this set, there can be no generalization, but how can a machine using current methods come up with this shared representation? How does it know what representation to use?

The water jar problem and many others that we have explored are fairly simple. They are limited not by the number of possible moves as was the problem with games like go. Rather, they suffer from what is so far a fundamental limitation—current computer systems cannot design their own appropriate problem representations. This is probably the central problem to having artificial general intelligence.

Insight

Insight is an essential part of intelligence. Recognizing that some known problem solution can be applied to a new problem is a necessary part of that insight, but I doubt that any current conventional approach to machine intelligence is capable of achieving that task. Metalearning would allow the system to learn about the capabilities of each of the known problem-solving approaches. At least in theory, metalearning would allow a system to select among the known solutions, but it is not adequate to create new solution spaces.

People may also have problems creating representations, but some of them do create them sometimes. To create an artificial general intelligence, we will need to find out how people create new representations and will have to create something analogous for machines.

The mathematician Henri Poincaré tried to describe this kind of problem-solving process from his own work:

> To invent, I have said, is to choose [among all of the possible variations in a given area]; but the word is perhaps not wholly exact. It makes one think of a purchaser before whom are displayed a large number of samples, and who examines them, one after the other, to make a choice. Here [in mathematics] the samples would be so numerous that a whole lifetime would not suffice to examine them. This is not the actual state of things. The sterile combinations do not even present themselves to the mind of the inventor. Never in the field of his consciousness do combinations appear that are not really useful, except some that he rejects but which have to some extent the characteristics of useful combinations. All goes on as if the inventor were an examiner for the second [academic] degree who would only have to question the candidates who had passed a previous examination. [From "The Foundations of Science" by Henri Poincaré, first published in Paris in 1908 and here translated from the French by G. B. Halstead.]

If invention were just a matter of choosing among known representations (Poincaré's "all of the possible variations"), then it would be amenable to current computational approaches. Invention would be nothing but search. But Poincaré goes on to dismiss this interpretation, noting that only some possibilities are considered. He is not at all clear about how some come to be selected for consideration, but that would be an important part of addressing this problem. Just as the expert chess player is selective about which potential moves she considers, an inventing intelligence has to be selective about just which possibilities are considered.

Poincaré goes on to describe some specific problems that he worked on:

> For fifteen days I strove to prove that there could not be any functions like those I have since called Fuchsian functions. I was then very ignorant; every day I seated myself at my work table, stayed an hour or two, tried a great number of combinations and reached no results. One evening, contrary to my custom, I drank black coffee and could not sleep. Ideas rose in crowds; I felt them collide until pairs interlocked, so to speak, making a stable combination. By the next morning I had established the existence of a class of Fuchsian functions, those which came from the hypergeometric series; I had only to write out the results, which took but a few hours.

Once he found a representation, verifying it took little effort and appeared almost automatic.

> Then I wanted to represent these functions by the quotient of two series; this idea was perfectly conscious and deliberate, the analogy with elliptic functions guided me. I asked myself what properties these series must have if they existed, and I succeeded without difficulty in forming the series I have called theta-Fuchsian.

He does not say in this essay why he wanted to represent these functions by a quotient, but that, too, gets pretty directly at identifying representations to solve problems.

> Just at this time I left Caen, where I was then living, to go on a geologic excursion under the auspices of the school of mines. The changes of travel made me forget my mathematical work. Having reached Coutances, we entered an omnibus to go some place or other. At the moment when I put my foot on the step the idea came to me, without anything in my former thoughts seeming to have paved the way for it, that the transformations I had used to define the Fuchsian functions were identical with those of non-Euclidean geometry. I did not verify the idea; I should not have had time, as upon taking my seat in the omnibus, I went on with a conversation already commenced, but I felt a perfect certainty. On my return to Caen, for conscience' sake, I verified the result at my leisure.

Poincaré then goes on to discuss two more examples of ideas revealing themselves to his conscious mind with little apparent immediate effort. In both of these cases, as well, he verified his results afterward when it was convenient.

Even though Poincaré could not describe explicitly the process by which he identified these novel representations, clearly some kind of work was going on in his brain. He noted that his new representations were analogous to some previous ones that he knew about; they were not invented out of whole cloth, but the method by which the analogy was identified remains unclear. Nevertheless, a method for selecting targets for potential analogies needs to be discovered if we are to build an artificial general intelligence. Perhaps we need it to take more excursions to the countryside.

One way to interpret Poincaré's observation is that it combined the ill-formed, call it intuitive, function of the unaided brain with the deliberate artificial intelligence of a trained mathematician. The natural intuitive capabilities of the brain operated to select ideas for consideration, perhaps on the basis of similarity. Both functioning together allowed Poincaré to identify solutions to his problems and then to deliberately verify them at his leisure.

It was only after he stopped deliberate thought about the subject that the intuitive native problem-solving could appear, a process called incubation. Many insight problems are solved, only after deliberate thought has been stopped, by a trip to the countryside or by a doze by the fire (Kekulé's dream discovery of the benzene ring). The convergence of natural intelligence and artificial intelligence was what apparently gave rise to these insights.

For computational intelligence to really be general, it will have to find a way to better emulate the insightful part of the human mind or find a way to substitute for it. Table 7 summarizes some of the features of natural and artificial intelligence in people.

Artificial intelligence is consistent with the intellectual accomplishments that we usually think of as intelligence. But the properties of what I am calling natural intelligence also play a role in human accomplishments. In cognitive psychology, these so-called natural properties are sometimes dismissed as bias or errors, but they also seem to play an important role in everyday cognition and in creative problem solving. A system that can quickly come to a conclusion before all of the evidence is in will be able to act more quickly than one that conducts a full analysis, but sometimes that quick conclusion will be wrong.

Natural intelligence depends strongly on pattern recognition. Things that seem familiar, in general, will not have to be deeply considered each time. Natural intelligence uses heuristics to guide its decisions, to limit the amount of processing it has to do. Again, these heuristics are not guaranteed to be correct. They are risky, but they may also be necessary.

Table 7

Natural Intelligence	Artificial Intelligence
Pattern recognition	Logic
Automatic	Deliberate
Fast	Slow
Incomplete	Complete
Lookup	Compute
Disorderly	Orderly
Jumps to conclusions	Reasons systematically
Inconsistent	Consistent
Impressionistic	Evaluative
Heuristic	Algorithmic
Implicit	Explicit
Diffuse	Focused
Associative	Statistical
Metaphoric/Analogical	
Emotional	
Impulsive	
Overconfident	

People invented artificial intelligence to overcome the shortcomings of natural intelligence. Artificial intelligence allows people to be systematic about their decision making. They may not evaluate all possibilities—even under the best of circumstances human rationality is bounded—but we have developed tools that help us to keep track of more alternatives than our natural intelligence can handle.

Computational intelligence typically requires many training examples for the system to learn. Humans can learn many concepts after only one or two examples. If we are to achieve general artificial intelligence in computers, we will need to reduce the effort required to achieve it. Requiring 20 million frames of experience to learn a video game is not a good long-term solution, no matter how fast we make a computer. Slow and deliberate learning could conceivably work for formal problems, like games, but it is entirely infeasible as a strategy for a dynamic physical world where not getting it right could be fatal.

Faster computers without better methods will not be enough. We need computer programs that can figure their way around problems, not just push through the designed space.

A Sketch of Artificial General Intelligence

Albert Einstein is quoted as saying that everything should be made as simple as possible, but no simpler. What computational intelligence has delivered has often been dramatic, but its narrow focus actually interferes with its ability to deliver artificial general intelligence. Current approaches to artificial intelligence may have oversimplified general intelligence out of existence by focusing on a small group of problem types (for example, games and other well-formed problems) and a narrow set of solutions to them.

The focus on a small group of tasks means that other tasks, which may be theoretically more important, have been neglected. The focus has allowed investigators to avoid much of the complexity of the physical and social world by incorporating commonsense knowledge implicitly in the representational models that have been used (their TRICS). The focus has been so complete that investigators do not seem even to notice that there are other kinds of tasks that have not been addressed.

The fear that computers will soon be able to improve themselves uncontrollably is due in part to the lack of awareness of just how limited the current mechanisms of computational intelligence are and the degree to which their capabilities are strongly dependent on their designers' implicit incorporation of common sense in the basic design. Current systems lack completely any common sense beyond that installed implicitly by their designers. Using current approaches, a computational intelligence program is no more capable of explosive self-improvement than a bumblebee is capable of reciting one of Shakespeare's sonnets.

Current approaches to artificial intelligence work by using data to adjust the parameters of a model that has been provided by the system's designers. These approaches excel at solving problems that can be addressed with this parameter-adjustment kind of process. System designers impose their common sense on the structure of these models, but so far, the computers themselves are not able to access any common sense that has not been imposed by the designer.

The recent progress in computational intelligence has come from the improved insight of designers to better structure the systems' representations, including the invention or discovery of heuristics that can limit the complexity of the parameter-adjustment process. Progress also comes from improved sources of data. Massive amounts of more-or-less curated data have become available from sensors, social networks, and other applications that were never available before. Continued improvement in computational intelligence depends on the combination of better representations and better data, not directly on better programs.

Computers have not yet achieved artificial general intelligence not because they lack some ineffable property that humans have, such as consciousness, but because computer scientists have not been designing for general intelligence. Solving the problem of artificial general intelligence will require designs that are capable of autonomous insight.

It seems very likely that artificial general intelligence requires a convergence between natural intelligence and artificial intelligence. Natural intelligence is what people do automatically, or easily, principally without explicit training. These are tasks on which computers have had only limited success. Artificial intelligence is invented. It is what people do with deliberate effort and explicit training. It is what computers do "naturally."

People's natural intelligence, which tends to be biased, incomplete, and approximate, but insightful and imaginative, needs to be combined with the logic and computational capabilities of computers because general intelligence demands all of these talents in order to be general. Too much rigor leaves an agent lost in thought, whereas too little rigor just leaves the agent lost.

Research will need to focus on these things, plus a number of others, to achieve artificial general intelligence. An artificial general intelligence agent will need to:

- Address ill-defined problems as well as well-formed problems.
- Find or create solutions to insight problems.
- Create representations of situations and models. What do the inputs look like; how is the problem solution structured (modeled)? What is the appropriate output of the system?
- Exploit nonmonotonic logic, allowing contradictions and exceptions.
- Specify its own goals, perhaps in the context of some overarching long-range goal.

- Transfer learning from one situation to another and recognize when the transfer is interfering with the performance of the second task.
- Utilize model-based similarity. Similarity is not just a feature-by-feature comparison but depends on the context in which the judgment is being conducted.
- Compare models. An intelligent agent has to be able to compare the model that it is optimizing with other potential models (representations) that might address the same problem.
- Manage analogies. It must manage analogies to select the ones that are appropriate and to identify the properties of the analogs that are relevant.
- Resolve ambiguity. Situations and even words can be extremely ambiguous.
- Make risky predictions.
- Reconceptualize, reparamaterize, and revise rules and models.
- Recognize patterns in data.
- Use heuristics even if their efficacy cannot be proven.
- Extract overarching principles.
- Employ cognitive biases. Although they can lead to incorrect conclusions, they are often helpful heuristics.
- Exploit serial learning with positive transfer and without catastrophic forgetting.
- Create new tasks.
- Create and exploit commonsense knowledge beyond what is specified explicitly in the problem description. Commonsense knowledge will require the use of new nonmonotonic representations.

The problems that are best known for demonstrating high levels of human intelligence, such as Nobel Prize–winning scientific insights, transcend the formal problem-solving approach typical of today's computational intelligence. They involve the formulation of new principles and, above all, new ways of representing their subject matter in the world. We will have to figure out how to achieve these tasks if we are to have a hope of generating an artificial general intelligence. Without this change in perspective, we have essentially no chance of ever achieving artificial general intelligence, let alone the superintelligence that scares Bostrom and others.

I believe that with the right investments, we will be able to develop computer systems that are capable of the full panoply of human intelligence. We cannot limit ourselves to looking where the light is bright and the tasks are easy to evaluate.

At some point, these computational intelligences may be able to exceed the capability of human beings, but it won't be any kind of event horizon or intelligence explosion. Intelligence depends on content as well as or perhaps more than processing capacity. The need for content and the need for feedback will limit the speed of further developments.

If we fail to develop artificial general intelligence, our failure will not be, I think, a technological failure, but one of our own imagination.

Resources

Bostrom, N. (2014). *Superintelligence: Paths, dangers, strategies*. Oxford, UK: Oxford University Press.

Copeland, B. J., & Shagrir, O. (2019). The Church–Turing thesis: Logical limit or breachable barrier? *Communications of the ACM, 62*(1), 66–74.

Driscoll, L. N., Pettit, N., Minderer, M., Chettih, S. N., & Harvey, C. D. (2017). Dynamic reorganization of neuronal activity patterns in parietal cortex. *Cell, 170,* 986–999.e16.

Dvorsky, G. (2018). New brain preservation technique could be a path to mind uploading. Gizmodo. https://gizmodo.com/new-brain-preservation-technique-could-be-a-path-to-min-1823741147

Goldhill, O. (2017). Humans are born irrational, and that has made us better decision-makers. https://qz.com/922924/humans-werent-designed-to-be-rational-and-we-are-better-thinkers-for-it

Gould, S. J. (1978). Morton's ranking of races by cranial capacity. *Science, 200,* 503–509. https://pdfs.semanticscholar.org/7992/a09d112b464fda63a8cae2859877cc2e0cde.pdf

Harvard Medical School. (2017). Neurons involved in learning, memory preservation less stable, more flexible than once thought. *ScienceDaily*. www.sciencedaily.com/releases/2017/08/170817122146.htm

Jabr, F. (2012). The connectome debate: Is mapping the mind of a worm worth it? *Scientific American*. https://www.scientificamerican.com/article/c-elegans-connectome

Kirkpatrick, J., Pascanu, R., Rabinowitz, N. C., Veness, J., Desjardins, G., Rusu, A. A., . . . Hadsell, R. (2017). Overcoming catastrophic forgetting in neural networks.

Proceedings of the National Academy of Sciences of the United States of America, 114, 3521–3526. https://www.pnas.org/content/114/13/3521.full; https://deepmind.com/blog/enabling-continual-learning-in-neural-networks

Kolbert, E. (2017). Why facts don't change our minds: New discoveries about the human mind show the limitations of reason. *The New Yorker.* https://www.newyorker.com/magazine/2017/02/27/why-facts-dont-change-our-minds

Lewis. J. E., DeGusta, D., Meyer, M. R., Monge, J. M., Mann, A. E., & Holloway, R. L. (2011). The mismeasure of science: Stephen Jay Gould versus Samuel George Morton on skulls and bias. *PLoS Biology, 9*(6): e1001071. doi:10.1371/journal.pbio.1001071

Loudenback T., & Jackson, A. (2018). The 10 most critical problems in the world, according to millennials. Business Insider. http://www.businessinsider.com/world-economic-forum-world-biggest-problems-concerning-millennials-2016-8/#2-large-scale-conflict-and-wars-385-9

May, R. M. (1976). Simple mathematical models with very complicated dynamics. *Nature 261,* 459–467.

McIntyre, R. L., & Fahy, G. M. (2015). Aldehyde-stabilized cryopreservation. *Cryobiology, 71,* 448–458.

Mullins, J. (2007). Checkers 'solved' after years of number crunching. *New Scientist.* https://www.newscientist.com/article/dn12296-checkers-solved-after-years-of-number-crunching

Pask, R. (2007). Checkers in a nutshell. http://www.usacheckers.com/checkersinanutshell.php

Pennachin, C., & Goertzel, B. (2007). Contemporary approaches to artificial general intelligence. In B. Goertzel & C. Pennachin (Eds.), *Artificial general intelligence* (pp. 1–30). Berlin, Germany: Springer.

Popova, M. (2013). French polymath Henri Poincaré on how creativity works. https://www.brainpickings.org/2013/08/15/henri-poincare-on-how-creativity-works

Sandberg, A., & Bostrom, N. (2008). Whole brain emulation: A roadmap (Technical Report No. 2008-3). Future of Humanity Institute, Oxford University. http://www.fhi.ox.ac.uk/brain-emulation-roadmap-report.pdf

Sardi, S. Vardi, R., Sheinin, A., Goldental, A., & Kanter, I. (2017). New types of experiments reveal that a neuron functions as multiple independent threshold units. *Scientific Reports, 7,* Article No. 18036. doi:10.1038/s41598-017-18363-1; https://www.nature.com/articles/s41598-017-18363-1

Schaeffer, J., Burch, N., Björnsson, Y., Kishimoto, A., Müller, M., Lake, R., . . . Sutphen, S. (2007). Checkers is solved. *Science, 317,* 1518–1522. doi:10.1126/

science.1144079; https://cs.nyu.edu/courses/spring13/CSCI-UA.0472-001/Checkers/checkers.solved.science.pdf

Schmidhuber, J. (1996). Gödel machines: Self-referential universal problem solvers making provably optimal self-improvements. IDSIA Technical Report. TR IDSIA-19-03, Version 5, December 2006, arXiv:cs.LO/0309048 v5; https://arxiv.org/pdf/cs/0309048.pdf

Schmidhuber, J. (2009). Ultimate cognition a la Gödel. *Cognitive Computing, 1,* 177–193.

Strathern, P. (2000). *Mendeleyev's Dream.* New York, NY: Penguin.

Sternberg, R. J. (1985). *Beyond IQ: A triarchic theory of human intelligence.* New York, NY: Cambridge University Press.

Turing, A. M. (1948). Intelligent machinery in mechanical intelligence. In D. C. Ince (Ed.), *Collected works of A. M. Turing* (pp. 107–127). Amsterdam: North Holland, 1992.

Turing, A. M. (1950). Computing machinery and intelligence. Mind, 59, 433-460.

Wixted, J. T., Squire, L. R., Jang, Y., Papesh, M. H., Goldinger, S. D., Kuhn, J. R., . . . Steinmetz P. N. (2014). Sparse and distributed coding of episodic memory in neurons in the human hippocampus. *Proceedings of the National Academy of Sciences of the United States of America, 111,* 9621–9626. doi:10.1073/pnas.1408365111

World Economic Forum. (2016). Global Shapers Annual Survey 2016. https://www.weforum.org/agenda/2016/08/millennials-uphold-ideals-of-global-citizenship-amid-concern-for-corruption-climate-change-and-lack-of-opportunity

Index